brave new worlds

Also by Bryan Appleyard:

The Culture Club: Crisis in the Arts

Richard Rogers: A Biography

The Pleasures of Peace: Art and Imagination in Postwar Britain

Understanding the Present: Science and the Soul of Modern Man

The First Church of the New Millennium: A Novel

brave new worlds

Staying Human in the Genetic Future

Bryan Appleyard

VIKING

VIKING
Published by the Penguin Group
Penguin Putnam Inc., 375 Hudson Street,
New York, New York 10014, U.S.A.
Penguin Books Ltd, 27 Wrights Lane,
London W8 5TZ, England
Penguin Books Australia Ltd, Ringwood,
Victoria, Australia
Penguin Books Canada Ltd, 10 Alcorn Avenue,
Toronto, Ontario, Canada M4V 3B2
Penguin Books (N. Z.) Ltd, 182–190 Wairau Road,
Auckland 10, New Zealand

Penguin Books Ltd, Registered Offices:
Harmondsworth, Middlesex, England

First published in 1998 by Viking Penguin,
a member of Penguin Putnam Inc.

1 3 5 7 9 10 8 6 4 2
Copyright © Bryan Appleyard, 1998
All rights reserved

Grateful acknowledgment is made for permission to reprint
excerpts from the following copyrighted works:
"Self-Portrait in a Convex Mirror" from *Self-Portrait in a Convex Mirror*
by John Ashbery. Copyright © 1974 by John Ashbery. Used by permission
of Viking Penguin, a division of Penguin Books USA Inc.
The Sense of Reality by Isaiah Berlin (Chatto & Windus, London).
By permission of Curtis Brown, London.
Sex, Art and American Culture by Camille Paglia.
By permission of Random House, Inc.

LIBRARY OF CONGRESS CATALOGING IN PUBLICATION DATA

Appleyard, Bryan.
Brave new worlds: staying human in the genetic future/Bryan Appleyard.
p. cm.
Includes bibliographical references and index.
ISBN 0 670 86989 9 (alk. paper)
1. Genetic engineering. 2. Genetic enginering—Moral and ethical aspects. I. Title.
QH442.A67 1998
174'.25—dc21   98-10243

This book is printed on acid-free paper.

Printed in the United States of America
Set in Janson
Designed by Mark Melnick

There may be in the cup
A spider steep'd, and one may drink, depart,
And yet partake no venom, for his knowledge
Is not infected; but if one present
The abhorr'd ingredient to his eye, make known
How he hath drunk, he cracks his gorge, his sides
With violent hefts. I have drunk, and seen the spider.

—William Shakespeare, *The Winter's Tale*

contents

acknowledgments

I would like to thank the following for granting me interviews in connection with this book: David Alton, Michael Blaese, David Botstein, Francis Collins, Kay Davies, Deborah Denno, Troy Duster, Peter Goodfellow, Tim Harris, Dale Kaiser, Philip Kitcher, Eric Lander, Victor A. McKusick, Matt Ridley, Alan Smith, Colin Tudge, and James Watson. I would also like to thank Daniel Dennett, Jared Diamond, and Lee Silver for interviews conducted in another context which turned out to be enormously helpful in this context. Particular thanks go to Peter Goodfellow for reading and commenting on the manuscript. Nigel Andrew, Anthony O'Hear, and George Walden helped me focus my thoughts, and my wife, Christena, and my daughter, Charlotte, helped me focus my life. My niece, Fiona Appleyard, about whom you will read in chapter 6, provided me with my most immediate awareness of the reality of these issues.

The Secret of Life?

Francis winged into The Eagle to tell everyone within
hearing distance that we had found the secret of life.
—JAMES WATSON, *The Double Helix*

The Eagle is a pub in Cambridge, England. I used to drink there when
I was a student at nearby King's College. Sometimes, after a few beers,
I would also tell everyone within hearing distance that I had found the
secret of life. But, by the next morning, I would always have forgotten
or abandoned the idea. I doubt that much was lost to the world. A collection of all the secrets of life discovered by every student who ever
drank beer would be unlikely to add anything to the sum of human wisdom. And anyway, students grow up, most of them deciding that there
is no such thing as the secret of life, or, if there is, that they are unlikely
to have any special knowledge of whatever it might be.

But when Francis Crick, a brilliant thirty-six-year-old scientist with a
slightly eccentric career record, "winged" into the Eagle in March
1953, he did have special knowledge, and it was knowledge that would
neither evaporate during the next morning's hangover nor vanish in the
cold light of adulthood. Yet it would have seemed to the other lunch-
time drinkers in the Eagle a rather obscure kind of knowledge. For all
Crick knew was how an extremely large but apparently very boring
molecule, found in the nucleus of living cells, was put together. Nevertheless, he thought he had discovered the secret of life. The world has

1

since decided he was probably right, for Crick and his American colleague, James Watson, had deciphered the structure of the molecule of deoxyribonucleic acid—DNA.

That structure—a double helix, a spiraling ladder, a twin coil—has since become the most familiar scientific image of our time. It has replaced Newton's apple, Einstein's hair, the "planetary" model of the atom, and the mushroom cloud of the hydrogen bomb. The double helix of DNA decorates books, posters, and almost every newspaper article or television program with any reference, however remote, to genetics or biology. Its simple, open-ended beauty has infected world culture. The image has become a fantastically condensed shorthand sign for "science," "life," "the future," and a whole range of other more ambiguous, often more troubling, meanings attached to the word "genetics." Few images now bear the global weight and significance of the twin spiral that Francis Crick bore in his head that day at the Eagle.

Many now say that Watson's and Crick's solution to the three-dimensional, sculptural puzzle of DNA was the most important discovery in human history. I think this must be philosophically true. I can think of no scientific insight that has so radically changed our view of ourselves. As I sit here writing this, I know that almost every cell in my body contains that double helix, as does almost every bacteria, virus, plant, and animal with which I share the world. Thanks to DNA, we know that all living things on the planet are related and that, at an unimaginably distant moment in the past, we were all born together as one replicating molecule in the cooling swamp of the primordial earth. No other human insight can hope to compete with this vision of the precise chemical connectedness of all life.

Yet, practically, it is still too early to assess the implications. It will take perhaps another fifty or a hundred years before we know how much our understanding of DNA can transform the world. Will it, for example, start a revolution more radical than the one that sprang from the cosmic mechanics of Isaac Newton, inspired, so the story goes, by an apple that fell in a Lincolnshire orchard more than three hundred years ago? Or will it enhance human capabilities as spectacularly as did the quantum theory born in the mind of Max Planck on a walk in Berlin's Grunewald woods in 1900? All we can say is that, at this moment, it seems almost certain that our ability to manipulate this huge but hypnotically simple molecule will leave nothing, including ourselves, unchanged.

Thanks to DNA, the science of genetics is now so vast and yet so intimate in its implications that it seems impossible to limit the ambitions of a book on the subject. Genetics now reaches outward to the universe and inward to our selves. Genetics is not just another technological development like the advent of antibiotics, computers, or space shuttles. It is not simply another phase in the spectacular progress of science marked by such theories as Darwinian evolution, quantum mechanics, or relativity. Nor is it merely a big idea like capitalism, communism, democracy, or facism that changes people's views of the world. Rather, it is all of these things—a historically unique combination of philosophy, science, and technology that confronts humanity with the most fundamental questions, our answers to which will determine the human future.

Genetics today places us at a vital moment in human history when we can choose not just how we are going to live, but who we are going to be. It is a big, global moment, but it is also an intimate, human moment. It affects not only politics, economics, and ethics, but equally the most private aspects of our lives. Historically it is unprecedented.

Yet there is a danger that people will sleep through this moment, only to wake and find themselves in a brave new genetic world. They will sleep because they feel excluded from the realm of science. They feel they have nothing to say about what scientists do and how they change the world. Thus they will slip into that most risky of modern habits—leaving science to the scientists.

This is understandable. Lay people have little choice but to regard science as a vast black box to which they can have no access. Occasionally the box emits "good news" in the form of new medicines or machines. But why and how it happened to be *this* medicine or *this* machine seem to lie beyond our understanding. Science has become such a large and complex enterprise that even scientists are obliged to take the authority of their colleagues in other disciplines on trust. For the rest of us the difficulty is compounded by the way science is sealed off from the everyday world by strange concepts and alien languages. The best we can do is read a few of the hundreds of popular science books that flood the market, in the hope of grasping, at best, a handful of scientific developments. But, of course, most of those books are written by scientists. They are immensely valuable—some of them, indeed, form an important category of modern literature—but they embody certain preconceptions. Most commonly, they are written from a position of power, acting as transmitters of carefully edited and simplified information to an audience that is excluded from the inner workings of the black box. This is not always made clear to the reader, and, as a result, the books tend to deepen rather than bridge the divide between scientists and the rest of us.

Let me use quantum theory as an example of this dilemma. Few people have any real grasp of even basic mathematics and fewer still can understand the advanced mathematics involved in quantum theory. Yet this theory is, for the moment, the most practically effective scientific insight ever. It gives us computers, radios, television, and the whole worldwide network of electronics that defines so much of the way we live and work today. It also provides us with intriguing clues to the ultimate nature of the physical universe, and when—or if—it is finally fully combined with the theory of relativity, it promises the tantalizing but disturbing possibility of a Theory of Everything, a conclusive set of equations that will embody the entire history of the universe. This will be a theory that will end three thousand years of fundamental physics.

Quantum theory began quietly and privately in 1900 when Max Planck took his walk in the woods. It continues, for the moment, almost a hundred years later with the electronically connected world in which we now live. Between the mind of Planck, meditating on some abstruse problems of physics, and the teenage Internet surfer is a direct and unbreakable link. On his walk, Planck had a vital insight into the workings of subatomic matter, while at the keyboard, the teenager plays fabulous games and routinely achieves feats of communication made possible by the exploitation of Planck's insight. Maybe this teenager does not know what Planck knew, but he certainly lives through and within that knowledge. It is a central determinant of his awareness of the world.

Genetics carries within itself the potential to change far more than quantum theory. It has already changed far more than we realize. And further changes will occur not simply through the creation of various useful gadgets, but rather through changes in the human self. The big

problem with such changes is that by definition, they will be very difficult to recognize. After all, the self that has been changed is the self
that is trying to understand that change. All the more reason, therefore, to understand—and argue about—genetics now, before it is too
late, and before our selves are irrevocably changed.

I know that many scientists will be impatient with this approach.
Some will feel it compromises their authority, and some may even be
insulted by the concept of lay understanding. Many will fear that I am
encouraging violent forces of bigotry and intolerance, that I am attempting to inaugurate a new irrationality—a new Dark Ages. I do not
underestimate the passions involved; within science there have been
deep and angry arguments about the future of genetics, specifically
about its political implications. Outside science, religious leaders have
forecast that reproductive issues arising from genetics will become a
more serious political issue than abortion. Since the abortion issue has
led directly to murder on the streets outside abortion clinics, nobody
should fool himself into thinking that these are trivial or remote matters only to be discussed at conferences or in seminar rooms.

I do not think we have any choice but to ask difficult questions of the
scientists. The alternative is to sit back and be grateful for whatever scientists choose to do, however their work may change the world. Science writers have a bad habit of accepting this state of affairs. For
example, two *Wall Street Journal* writers, Jerry E. Bishop and Michael
Waldholz, blandly assert in their book *Genome*—in the midst of a thor-
ough history of the Human Genome Project—that "social mores
change rapidly in the face of new technology." This may be true, but it
is just too big a point to be made in passing. One wants to halt there
and, before reading further, demand another book examining the implications. For, if social mores do change so rapidly, then that must condition our every response to the scientific wonders the authors describe
and must change the role of popular science books. Without an awareness of the critical interface between science, ourselves, and our society,
science writing becomes an essentially meaningless catalogue of events
or, worse, a vacuous fan letter.

In short, it is wholly unrealistic for science to claim for itself a kind of
untouchable innocence about its impact on the world. Scientists themselves do not pretend to preserve that innocence, and it is absurd to ex-

pect the rest of us to accept it. We have every right—indeed, we have an obligation—to argue with science and the scientists. As Senator Edward Kennedy said in the mid-seventies, when anxieties about the new power of biology were first beginning to surface: "There can be no turning back to the days when scientists were left totally on their own to chart their own course. . . . The public will immerse itself in the affairs of science. Whether it does so constructively will depend on the willingness of scientists to welcome public participation."

Of course, there are many scientists whose work reinforces their religious faith. Francis Collins is a devout Christian and director of the National Center for Human Genome Research at the National Institutes of Health in Bethesda, Maryland. He is, therefore, at the very center of modern genetics, and once told me: "God isn't afraid of science. He knows this stuff. And the notion of not pursuing genetic research is unthinkable. It contains within it the seeds of the alleviation of much human suffering. It would be unethical to delay it. Christ was a healer."

On the other hand, James Watson, codiscoverer with Francis Crick of the structure of DNA and now president of the Cold Spring Harbor Laboratory on Long Island in New York, made it clear to me that both he and Crick were interested in the ultimate molecular constituents of life precisely because such knowledge would represent a challenge to religion; it would show the strictly material basis of life. Clearly, this was not the only reason for their project, but it was the style of their thought.

Collins is a Christian, Watson an atheist. Through his microscope, Collins sees the work of God; through his, Watson sees only the incredible intricacy generated by 3.5 billion years of blind evolution. As scientists speaking a common scientific language, they both see the same things—cells, chromosomes, the slender, double helical thread of DNA. As men, they see something utterly different—so different that it seems impossible that two people could have so much and yet so little in common. Somewhere in that gap between those two brilliant men who agree on everything and yet nothing lies this book.

I suppose one form of the question I am asking is this: If knowledge is not morally neutral, can it then be bad? Is it possible that we can actually learn something that contradicts our most humane instincts?

Have we already learned it? Clearly I hope not, but, equally clearly, I am not encouraged by the way we currently talk about genetics, the most restless, turbulent, and demanding form of knowledge that our species has yet produced.

I can easily picture Francis Crick "winging" into the Eagle. The pub did not change much in the sixteen years between that moment and my arrival in Cambridge. It was still a dark, smoky cave promising intoxication and revelation. And, of course, the ghost of the moment still lingered in the walls—for how could such a powerful spirit ever be laid to rest? It was the first time in history that a man, with all the authority of modern science, could seriously claim to have found the secret of life. Imagining that moment provokes an ominous yet thrilling shiver at the thought, the awful possibility, that Francis Crick's triumphant announcement in that Cambridge pub may have been the truth; that he may, in his beautiful, spiraling, molecular model, have seen the secret of life. What on Earth might that mean?

chapter 1

The Future

The Future

Biology is likely to dominate science for the next century.
—KAY DAVIES, professor of genetics, University of Oxford

Forecasting the future is dangerous. Sometimes we expect too little to happen. At the end of the nineteenth century, some physicists believed their subject was almost complete. This was just before Planck's quantum theory and Einstein's relativity theory started a revolution in physics that was to dominate the next fifty years of science, posing fundamental problems that remain unresolved.

Or we expect too much. According to the 1968 film *2001: A Space Odyssey*, we should by now have self-conscious computers and be ready to send astronauts to Jupiter. In fact, since the moon landings in the 1960s, humans have not even left Earth's orbit and artificial intelligence research has been stalled for twenty years.

The difficulty is that any forecast assumes that present developments will continue on a straight line into the future. As a result, we make naive assumptions. But how else are we to proceed? It is, by definition, impossible to forecast the unforeseen. We have only the present as evidence, so the futures we construct are always destined to be commentaries on the present. The reality is that problems and breakthroughs, not straight lines, are the main determinants of progress. The cost and dubious benefits of space travel have stopped us going to Jupiter or

———— 8 ————

even returning to the moon, while quantum theory has produced an information revolution. Neither the problem nor the breakthrough could have been anticipated.

Forecasting the future of genetics is a riskier endeavor than most. At one level, the future looks very predictable indeed: The overall pattern of developments in molecular biology seems clear and we can, therefore, expect more of the same. Most geneticists seem to know exactly where they are going. Yet at another level, it seems impossible to predict anything. Our present genetic knowledge is notably lacking in direct, practical applications. We may be able to continue increasing our knowledge of what genes do, but we do not know what to do with that information, or, indeed, whether we can do anything. Some say molecular biology is still waiting for its Newton or Einstein, its big theorist. In this view, the unforeseen element is the overarching theory that makes sense of it.

One final and, for geneticists, depressing possibility is that genetics has distorted the entire discipline of biology. Maybe we have taken DNA too seriously. Perhaps proteins are, in reality, the most decisive substances in the biochemical system. They, not DNA or its precursor replicating molecules, may have been the start of the life process. By concentrating so much on the gene, we may have ignored the dynamics of the whole organism, and once we escape from this delusion, genetics will decline in importance, becoming one aspect of biology.

Whichever future we choose, we are likely to be wrong. Yet the present implications of developments in genetics are too vast to be ignored. The future presses in on us daily. News stories demand that we ask ourselves, What if? Dolly the sheep stunned both the scientific and the lay world in 1997: Cloning from adult cells was now possible. Suddenly the future seemed to have arrived. But not quite, for such stories are all about developments as yet unapplied in the human realm, and they can only be understood from the perspective of a possible future.

Scientists tend to be reticent about addressing this. They dislike sensational stories about the possibility of "designer babies" or "brave new worlds" of cloned human beings. They say either that we are a long way from being able to do anything like *that*, or that there is no reason to do it, so it will not be done. But Dolly undermined the credibility of the first position. (The day before she was revealed to the world, it would have been easy to find a large number of geneticists who would have insisted that cloning from adult cells was either impossible or a long way in the future. All of them would have been wrong.) And the widespread insistence that there could be no reason to clone human beings simply betrays an innocence about the infinite demands of human vanity. This is the age of consumption—if it can be bought, it will be.

So we must look forward. Perhaps the most low-risk forecast concerns the Human Genome Project—the reading of all the chemical information stored in human DNA. Early in the next century, "Adam II" will have been created. We will have a complete list of the three billion chemical letters that compose the message of the human genome.

From that point onward forecasting becomes much more risky because we do not know how we will be able to apply that information or, indeed, how much of it we shall understand. It could be many years before we can interpret all the messages buried in the genome. And it may take decades or centuries to work out the complete biochemistry of the body. At that point we shall be able to say that the study of anatomy is complete—that we shall know the entire system of the human body. Perhaps, some have suggested, this stage will be reached around the year 2100.

It is generally assumed that both before and after the completion of the Human Genome Project, applications will flow continuously from the information it provides. These applications will be primarily about the improvement of the physical human condition. Medically, this is an urgent matter because we seem to have almost reached the limit of improvement using conventional methods. Doctors hope genetics will lead them away from a road that is looking increasingly like a dead end.

For example, life expectancy has almost doubled in the developed world over the last hundred years, creating hopes among many that this rate of improvement will continue (straight-line thinking again). But, in fact, for those reaching the age of sixty, life expectancy is only a few months more than it was at the beginning of the century. This suggests that our main achievement has been to stop people from dying young rather than to make them live longer. We have failed, so far, to do anything about aging other than to allow more people to experience it.

Genetics has provided many speculative possibilities for extending the human life span by slowing, or compensating for, the changes that take place at the molecular level. Aging may be, after all, an entirely genetic process caused by a buildup of mutations in the DNA of the somatic cells. Or, even if aging is not directly genetic, it may involve cellular processes that can be genetically regulated.

The difficulty is that any attempt to intervene entails the thwarting of nature at a very fundamental level. From the point of view of the gene, our longevity is neither here nor there. Once we have finished reproducing, it is immaterial to our genes whether we live or die; we have done our job of transmitting them to the next generation. Yet even so, there is biological evidence that death does not have to happen. Stem cells in the bone marrow and cancer cells are immortal; they are capable of reproducing themselves forever. So, at the cellular level, death is not an inevitable outcome for biological systems. If we can discover what keeps those cells going, then perhaps we can treat all other cells to behave in the same way. Immortality is no longer completely unthinkable.

Currently, eighty-five might be said to be the optimum human age— the life span which a reasonably healthy, and lucky, person might expect. The utopian vision of current Western medicine is that of a person who drops painlessly dead on the golf course at eighty-five after a disease-free life. However, some now talk of using genetics to increase that optimum age to one hundred twenty-five. Yet even if these scientists were successful, the problems, both for the individual and society, would be enormous. For example: How much of this increased life span would be spent in helpless dependency?

Oddly enough, the conquest of individual diseases does not have much effect on the statistics of longevity. Almost all the increase in life span in this century has been due to public health measures and improved nutrition. Even curing cancer would add only a year or two to the average Western life. Yet, clearly, it is better not to be ill, and, to the individual with cancer, statistics provide little consolation. So it is the treatment of illness that is the primary driving force of genetics.

The immediate future of medical genetics looks reasonably clear. The 4,000 genes involved in single-gene disorders will be progressively identified, and scientists will attempt to identify the genes involved in polygenic illnesses—those arising from the interaction of a number of genes. This is a much more complex process, but it will bring genetics into many more lives since single-gene disorders are rare, while polygenic disorders are common.

A single-gene disorder like Huntington's disease or muscular dystrophy tends to be a simple, all-or-nothing event: If you have the single dominant gene for Huntington's or the two recessive genes for muscular dystrophy, you will get the disease. Some environmental (not directly genetic) factors may determine the time and rate of onset, but these do not alter the fact that the conditions are genetically predestined. This makes tracking the genes with the current state of knowledge and technology relatively simple—they follow clear Mendelian patterns of inheritance with no significant ambiguity about the role of the genes as opposed to the environment. Here is a bad gene and it causes this disease.

There remains, however, the ambiguity about the evolutionary significance of these aberrant genes. Tampering even with these apparently clear-cut cases of disease genes may not be a simple matter. Genes that do one thing may have other, entirely unrelated effects—a phenomenon known as pleiotropy. Genes that appear to be bad may provide benefits of which we know nothing. A single sickle-cell anemia gene, for example, provides some protection against malaria. To describe a gene as "bad," therefore, is risky—it may be "good" in ways we cannot yet imagine. But, leaving that ambiguity aside, the origins and broad causal pathways of many single-gene disorders are now understood, even though little or nothing of therapeutic value can yet be done with that understanding.

Multigene disorders, however, are much more complex. Even if just two genes are involved, the difficulty of tracking them increases enormously. There are three billion chemical bases and maybe 100,000 genes, all of different shapes and sizes, in the human genome. As many as 20 genes may be involved in some disorders. Even if we were to successfully track these genes, the picture would be far more ambiguous than that of the single-gene disease. Say we were able to identify a number of genes linked to a form of heart disease. This would tell us something about the health prospects of anybody possessing that combination of genes, but it would not be the sort of clear-cut information we get when we identify single-gene disorders. It would say simply that this person is *at risk* from or *predisposed* to heart disease, not that he will definitely get it. If that person works out and doesn't drink, smoke, or eat too much fat, then he can reduce that risk and may become ill and die as a result of something entirely unrelated to heart disease. Or he may not.

Almost all polygenic disorders are likely to involve some degree of interaction with the environment. In the case of the man at risk from heart disease, "environment" refers to his fitness, eating, drinking, and smoking habits. Even if this environment is corrected to take account of the increased risk, the man's heart disease genes may interact with other genes and lead to a heart attack anyway, no matter what his general state of health.

The environment encompasses not only psychological and physiological factors, but also those more conventionally seen as "environmental": air quality, pollution, heat, cold, sunlight, and so on. The impact of some of these elements have been isolated—for example, tobacco and sunshine are factors in lung and skin cancers respectively—but many have not. In fact, it is probably misleading to think of the environment in which we live as simply benign or harmful. Of course, some elements of that environment, such as clean air, may be benign, and some, like tobacco smoke, may be harmful. But the picture that is now emerging is that life is a constant interaction with the external world at the molecular level. All life for as long as it manages to endure is an incredibly ingenious victory against huge environmental armies.

Disentangling the incidents in this war will be difficult and our conclusions may frequently be illusory. So the next wave of medical information arising from genetics is not going to be nearly as clear as the first wave. Medical knowledge will move into the realm of probabilities. For example, say that in ten years' time, a doctor takes a DNA sample from a fetus. He then sequences the DNA, using faster technology than anything we now have, and seeks out whatever single-gene or polygenic disorders have been identified by that time. He finds that the fetus has a combination of genes that makes it more prone to heart disease than the average person. The doctor will be able to quantify the risk. But the figures will not be simple: He may say, for example, that, given a normal lifestyle, this person will have a 46 percent chance of having a heart attack by the age of thirty. Given an unusually healthy lifestyle, the risk may fall to, say, 39 percent by the age of forty.

These figures will be based on large population studies which tell the doctor that, say, out of every hundred people with these genes who maintain a healthy lifestyle, 39 had heart attacks by the age of forty. This may sound like solid information—and for medical policymakers, insurance companies, or hospitals, it is, since they look at large numbers of people, whereas we only look at a few, generally ourselves and our families. But what are the parents to make of these figures? Does it mean their baby is ill? When it is born it will almost certainly look and behave like any other baby. Yet there is this strange phantom of probabilities, of calculable futures, already present at its birth. We have always known that our newborn babies will one day die, but soon we will be told when and why it is likely to happen.

This involves the whole relationship of the individual to a statistically described world, and it calls into question our concept of disease. It will demand more of patients and doctors and, most importantly, more of their relationships.

The potential for refinement of such predictive thinking is huge. Any illness that can be shown to have a genetic component—and that, some argue, means every illness—may be detectable as a probability or predisposition from a sample of our DNA. Vast probability charts— scientific horoscopes—of our susceptibilities may be constructed and whole lifestyles designed to minimize risk: This person should not smoke and that person should avoid red meat or taking a job in a particular kind of factory. In the future we may be given, at birth, a life-long program precisely tuned to our own genetic predispositions.

This need not mean we will be enslaved by statistics. It may in fact mean we will be liberated. For example, current medical advice for the prevention of heart disease includes cutting down on animal fats, exercising, and avoiding smoking. But in reality, doctors know perfectly well this advice is based on ignorance; they simply do not know whether any particular person needs to do those things, only that, *in general,* people do. If we knew which people's arteries were likely to become clogged as a result of the consumption of animal fats, then we could just tell them to cut down their intake. The rest of us could eat as much as we liked, and possibly some people could even smoke as much as they liked. In this sense genetics could free people from a whole range of anxieties that doctors feel obliged to impose on the whole population on the basis that they may improve the health of a few. Our health profiles would become much more precise and personal. In theory, the whole health-care business could become much more efficient.

And it may not simply be familiar physical illnesses that are drawn into this realm of probabilities and predictions. Most serious mental illnesses are now thought to have genetic components. Manic depression and schizophrenia are assumed to be overwhelmingly genetically determined, and a worldwide search is in progress for the genes involved. Other, less serious mental disturbances may also be genetically diagnosed.

At this point the question of what is and is not an illness arises. Schizophrenia is obviously a debilitating disorder, as is manic depression. But what about simple depression? Currently, how we view a particular state of depression may result in hospitalization, psychotherapy, a consoling chat, the prescription of antidepressants, or a brisk insistence that life is tough, pull yourself together, and get on with it. The dividing lines between the levels of seriousness will be vague. But genetics may offer more precise divisions. As one researcher put it to me: "[Genetics] provides an objective basis for a subjective diagnosis." We may be able to say that this depression has a recognizable genetic component and that one does not.

So, in looking to the future, the question becomes: How much of human life will be explicable in terms of genetics? In the current climate the usual answer is: Almost every aspect of human life has a large and frequently decisive genetic component.

The psychologist Thomas Bouchard has said, "For almost every behavioral trait so far investigated, from reaction time to religiosity, an important fraction of the variation among people turns out to be associated with genetic variation. This fact need no longer be subject to debate; rather it is time instead to consider its implications." This is the mainstream, working assumption in science today. On that assumption, schizophrenia, as well as other forms of behavior we would not necessarily consider as evidence of disease, could be explained by genetics. Routine depression, a tendency to lose one's temper, sexual promiscuity, or anything even slightly out of the ordinary could be shown to have genetic roots. In itself this is not surprising; we often notice that children inherit not just the looks but the attitudes of one or both of their parents. What will be new in the future will be, first, exact knowledge of the precise genes involved and, second, the possibility of doing something about such tendencies.

Of course, we might now say, But you are not sick if you tend to get angry. But what sickness is or is not tends to be defined by the prevailing wisdom. Just as genetics may come to teach us that somebody with a predisposition to heart disease later in life is sick now, so it may come to convince us that certain personality traits are sicknesses and should be treated as such. Look at the way surgery is now widely used for cosmetic purposes. Yet do we regard ugliness as a sickness? Many people, both doctors and patients alike, now act as though it is. The possibility of changing any human condition immediately transfers that condition into the medical realm—a place dominated by the simple polarity of sickness versus health.

This makes it highly unlikely that genetics will reduce overall medical costs, though many have hoped that this would happen because of the resulting increased effectiveness of preventive medicine. But if genetics also increases the number of human attributes classified as illnesses—or simply as treatable disadvantages—then medical spending will continue to rise. It will also tend to be even more concentrated on the richest sectors of society—those most able to afford the diagnostic and therapeutic programs involved.

But there is another form of "preventive medicine"—abortion. Knowing the gene for a particular disease may not yet lead to any treatment, but it does, if the gene is identified prenatally, offer the choice of abortion. Abortion is the one medical intervention based on medical genetics which is already widely available. Prenatal tests are offered to women thought to be at risk of having genetically abnormal babies. These tests—such as amniocentesis or chorionic villus sampling—are invasive and slightly risky, so they are not offered to every mother-to-be. If they are positive, abortion is frequently the only response available.

But if genetic flaws can be identified even earlier, abortion can sometimes be avoided for at-risk families. The use of preimplantation diagnosis, pioneered at the Hammersmith Hospital in London for families at risk of cystic fibrosis, involves taking eggs from the mother, fertilizing them in vitro with the husband's sperm, and incubating them for three days, still outside the body, until they form a cluster of eight cells. One cell is then removed from each embryo and tested for the disease gene (rather amazingly this removal of one eighth of the mass of the embryo has no effect on its future development). If the disease gene is present, the embryo is discarded. A healthy embryo is then reimplanted into the mother.

More rapid DNA sequencing techniques and greater knowledge about the effects of specific genes would mean that a much larger range of conditions could be sought in the embryonic cell. These conditions need not be what we would now classify as serious diseases. In time they could, for example, forecast anything from the eye color to the likely intelligence or sexual orientation of the child. Preimplantation diagnosis could offer, to those who could afford it, a choice of what kind of child they would like.

But prenatal testing on a larger scale will only happen when a cheaper, easier method is found of getting at embryonic or fetal cells. The best hope is the use of a simple blood sample taken from the mother. Fetal cells do circulate in the mother's blood and could, therefore, be isolated and tested. This would mean that all pregnant women could be routinely tested and, subsequently, offered a vast amount of information about their babies. However, much of this information would be in the form of statistical probabilities. Confronted with this kind of information, people will seek explanations. Thus, one absolute certainty for the future is that genetic counseling is bound to be a massively expanding industry.

Also in the near future a number of specific medical treatments are likely to become available. The overwhelming consensus among geneticists is that the development of highly specific though conventional treatments will be the first effective fruits of the Human Genome

Project. Because genetics will tell us a lot more about individuals and their biochemical constitutions, we will be able to target particular conditions more exactly. In this form, genetics can be seen as no more than an extension of existing medicine. We simply will know a great deal more about the way the body works and we will be able to provide treatments that look much the same as existing treatments (pills, injections) but that act more precisely not only on a specific disease but also on the specific individual. This process will be rapidly advanced by the genetic engineering of animals so that, from the point of view of the disease, they become more like humans.

"Mice don't get Alzheimer's," explained one man involved in this industry, "[and] they don't get high cholesterol. But we can genetically engineer these conditions into an animal so you have a whole animal system other than a human being on which to measure and evaluate drugs. This tool, before we had genetically engineered animals, was unavailable. It will accelerate the drug development process. Think how valuable that tool can be to any drug-testing or drug-development company."

Such treatments will also result in a massive expansion of the existing industry of protein production via genetic engineering. Currently this production of protein happens through either engineered bacteria or animals—typically goats and sheep—that have been genetically altered to express a human protein in their milk. Assuming this technology continues to be the most effective method, we can expect the development of huge new protein factories and farms, the latter probably relying on techniques developed once the mysteries of the cloning of Dolly have been deciphered.

Animals, usually pigs, will also be more radically engineered to produce organs suitable for transplantations to humans—so-called xeno-transplantation. This was first tried, absurdly prematurely, in 1964 in Jackson, Mississippi, when a baboon heart was transplanted to a human who, of course, died. Now the procedure is technologically possible but fraught with uncertainties about the precise tuning of the immune response and the possibility of transferring animal viruses to humans. This is exactly what appears to have happened in the case of acquired immune deficiency syndrome (AIDS), a disease which is thought to have found its way into humans from monkeys. The science of immunity seems certain to benefit from the knowledge derived from the genome project. A reasonable straight-line forecast would be that xenotransplantation will happen quite soon and will become commonplace over the next couple of decades.

Distinctively human body parts also may be produced if we succeed in mastering the control mechanisms of the genes. Again, Dolly seems to point the way by suggesting that the genes in differentiated cells may be controllable. For example, we might begin with one cell from a patient and, by switching on the right genes in the right order and in the right environment, grow a new arm or leg. In this case there would be fewer or no problems with rejection by the immune system because the new limb would be genetically identical to the rest of the patient's body.

Finally, knowledge of gene regulation may also lead to a breakthrough in the treatment of cancer—the disease which kills one in three of us. All cancer is genetic because it involves an alteration in a cell's DNA. Some types are also genetic in the sense that they are hereditary, but the remainder seem to be the result of mutations due to environmental factors or the action of viruses which replicate by changing the DNA of the host cells. If we can regulate genes with sufficient accuracy, we can, in theory at least, turn off cancer-causing genes or otherwise change the genetics of cancer cells to eliminate tumors.

Already we have isolated a number of genetic factors involved in cancers. The first breakthrough came as long ago as the 1960s when the American scientist Peyton Rous discovered a gene—christened "sarc"—that caused cancer in chickens. Since then, there have been steady though not spectacular developments. Colon cancer and breast cancer, the most common form of the disease in women, have been traced to certain genes, called oncogenes. There will be a huge increase in the number of cancer-causing genes we are able to identify.

Genetic advances have already transformed the way cancer is regarded. Traditionally cancers have been categorized on the basis of the part of the body in which they occur—colon cancer, lung cancer, and so on. It is now becoming clear that it is more accurate to categorize cancers on the basis of the chemical pathways involved. So a cancer of the throat may, biochemically, have more in common with a cancer of the bowel than it does with another throat cancer. As the geneticist Eric Lander said: "We will classify the tumor according to the pathway, not the site in the body."

One form of specific human cloning is currently being used as a treatment for cancer. This involves the production of so-called monoclonal antibodies. White blood cells are removed from a patient and exposed to cancer cells in the hope that they will produce the correct antibodies to destroy the cancer. Only about one in 100,000 white cells will react in this way. But this is enough if they can then be cloned and injected back into the patient to seek out and destroy tumors. This technique has long been promising, but results are so far inconclusive.

These are simply examples of the use of genetics against cancer. The field of cancer research is so vast that there are many others. Current cancer treatment is still largely based on the conventional methods of chemotherapy, radiation, and surgery. In spite of all the effort in the past half century, cancer treatment has only been refined, not revolutionized. While would-be revolutionaries now look to genetics, in view of the history, it would be safest to say only that some kind of breakthrough may happen in the next ten years.

Gene therapy, for the moment, remains a more distant prospect. "The theory of gene therapy is impeccable," one geneticist told me. Unfortunately, the practice is not. There is a gap between what happens in the test tube and what happens in patients. It has been suggested that it will be at least fifty years before gene therapy becomes a widespread medical practice. The difficulties of targeting genes and changing enough cells to have any effect have, so far, proved insuperable. "Gene delivery," as Michael Blaese at the National Institutes of Health in Bethesda put it, "is the issue."

The recent development of an artificial human chromosome suggests one way of delivering genes. In 1983, a yeast artificial chromosome (YAC) was created. YACs can now be made to order and are used to store millions of DNA base pairs. They have been described as the "beasts of burden" for the Human Genome Project. After that the race was on to make what was conveniently called the Big Mammalian Artificial Chromosome—the Big Mac. This has finally resulted in the creation of an artificial human chromosome.

The point about artificial chromosomes is that they may offer a way of carrying exactly the genes and regulatory sequences we need into any organism, so that transgenic animals—creatures which carry human genes—can be more efficiently created than they are at the moment. The genes would simply be ferried into their cells using the artificial chromosome.

Similarly, gene therapy could be made to work. In both cases new genes would be permanently added and would not disturb the work of existing genes. One danger, unfortunately, of existing gene therapy techniques is that, in inserting genes into the targeted place, we can end up inserting them in the wrong place and distorting healthy sequences.

Germ line therapy—in which the sex cells are altered so that future generations are treated by current medical interventions—is also on the agenda, but here the issues are more complex. It may seem obvious that if someday we are able to change the DNA in sex cells so that future generations will be immune to AIDS, then we should do so. As yet there is no prospect of that happening, but even if there were, could we ever be sure that we were not creating new problems for the future? If, for example, we could eliminate sickle-cell anemia through germ line therapy, would that result in an increase in malaria? Tinkering with the whole human gene pool may have consequences we cannot possibly foresee. Will we have so much confidence in our knowledge that we feel able to alter the outcome of millions of years of evolution?

The foregoing examples are only a sampling of possible future developments in medical genetics. Many others will be announced month by month, including some that will come as a complete shock, even to geneticists. The general picture, however, is clear: We are at a stage in our understanding of human genetics where much is promised but little of direct, practical value has been achieved. In fact, the word "much" is probably too weak. It is more true to say that *almost everything* is promised—from immortality to a cure for cancer to a new consumer market in offspring characteristics. Developments in genetics offer the possibility of bringing all our life processes under control.

Anything, it is now clear, may be possible. And that includes penetrating the human mind. Molecular biology in general and genetics in particular offer new ways of understanding the mind. Evolutionary psychology is now an active and fashionable area of research, based on the idea that our psychological condition can be investigated by viewing the mind as a product of evolution rather than as a blank sheet written over by culture and environment. We can now study human sexual behavior in the same way we might study that of chimpanzees—as a series of strategies that have been tried and tested by the evolutionary process and passed on by heredity. The central biological insight that we are connected to all other living things is being extended to the way we act and think. Psychology, psychiatry, sociology, and even politics can be viewed and analyzed as the products of evolution. In the future it is quite possible—indeed, to some extent it is already happening—that psychoanalysts will refer to Darwin rather than Freud when attempting to understand mental disorder. It is likely that, in time, the study of the mind will become much more like the study of anatomy.

Meanwhile, Francis Crick has forecast the explosive growth of "molecular psychology"—the study of the workings of the brain at the molecular level. "The present state of the brain sciences," he has written, "reminds me of the state of molecular biology and embryology in, say, the 1920s and 1930s. Many interesting things have been discovered, each year steady progress is being made on many fronts, but the major questions are still largely unanswered and are unlikely to be without new technologies and new ideas."

Molecular psychology could link up with evolutionary psychology and with research into genetic predisposition to mental disturbance to form a new science of the mind based entirely in biology. "We will have," as one researcher put it, "a much clearer understanding of what personality means biochemically."

But, in the future, genetics may also affect much more than just the individual human life. For a start, the medical developments I have noted have wider policy implications. Accurate prenatal or even postnatal diagnosis of future conditions may lead to a nightmare for insurance companies and for their customers. In countries like the United States, where the health care system is largely financed by private insurance companies rather than the government, as in Britain, a conflict exists between the commercial interests of the companies and the public interest in providing health care. If DNA tests for a wide range of diseases are demanded by insurance companies, then, inevitably, a large number of people are going to find themselves uninsurable or with cripplingly high premiums.

All the geneticists I have met have said—perhaps because of their political inclinations—that genetics will destroy the present American health care system. They say it will have to be replaced with some kind of national health service like Britain's. Whether this is true or not, it is clear that developments in genetics will force huge changes in the way health coverage is provided. The full implications of these changes are only just beginning to be felt.

Advances in genetics also have other, even darker, policy implications. First, they suggest that human characteristics are, to a greater or lesser extent, fixed at birth, that people are made by nature (their biology) rather than by nurture (their environment). Second, they provide information not just about individuals, but also about groups of people. For example, we may well be able to identify a whole range of precise differences between races, as well as genetic dispositions to criminality, dissidence, or any other traits perceived as socially undesirable. The readout of your genome at birth may not just tell your doctor how likely you are to have a heart attack; it may also tell him how likely you are to rob a bank or blow up a federal building. Genetic information may be able to create clearly definable—and very easily identifiable—criminal or politically dissident classes.

These are scientific developments which can only really be understood in moral, philosophical, and political terms. The improvements in the physical human condition that may flow from developments in genetics will be trivial compared to the catastrophe that will befall us if we mishandle the information in the wider public realm. Concealed within the knowledge we are now acquiring are insights that may be profoundly socially divisive and which could overthrow the basis on which the wealth and stability of Western democracies are constructed. Any forecast of the future must make one of two assumptions: Either we manage this deeper genetic knowledge wisely or we do not. In the first case we can be reasonably optimistic. In the second case there need be no limit to our pessimism.

Outside the human realm, advances have already resulted in direct, practical applications. The primary theme of developments in the genetics of animals and plants is the breaking down of barriers between species. Transgenic animals are now relatively commonplace, as are transgenic plants. These "chimeras" can be created because of our power—derived from the recombinant DNA technology developed in the early 1970s—to move DNA from one species to another. I have already mentioned the use of transgenics in medicine. In agriculture, transporting genes across species results in the enhancement or preservation of favored characteristics far more rapidly and efficiently than traditional breeding methods. Crops can be made resistant to disease or able to endure a much higher level of chemical treatment to eradicate destructive pests. Or they can be made to produce fruit with more convenient characteristics, like the now famous tomato that lasts longer on supermarket shelves (thanks to a gene christened the Flavor Saver Gene by Calgene, the company involved). Dozens of species have already been transformed, and, within a few years, techniques will become available for the genetic manipulation of all major crops species.

These advances could go much further than merely producing better versions of the same species; they could lead to wholly new species. The writer Colin Tudge speculates, for example, about the possibility of a Virginia creeper that does not respond to the seasons and is tolerant of cold, providing insulation for houses over the winter and producing strawberries. Eventually, Tudge speculates, we can move on to developing life from scratch: "So perhaps in a hundred years, and perhaps less, there will surely be a 'life creation project' comparable with the Human Genome Project of today."

The British geneticist Steve Jones takes this even further. In his view, plants could become biological factories, producing anything we want and completely displacing animals: "All this may mean that plants will soon do almost everything and that animals will fade in importance as—perhaps—the salmon-flavoured banana takes over." And he suggests: "The rural landscape may become one in which asexual cows feed on engineered grass under the shade of clonal trees."

And so on and so on. We are in the process of taking control of life. Some aspects of this process will appear to be no different from previous scientific and technological developments. Other aspects, however, will be profoundly different. We will produce new species, diagnose illness long before it happens, "know" human beings at the biochemical level, manipulate our reproductive processes, and change ourselves.

Such developments are like nothing that has gone before. They represent a fundamental redefinition of human capability. The remainder of this book is about what that really—as opposed to merely scientifically—means.

chapter 2

God, the Bomb, and the Double Helix

God, the Bomb, and the Double Helix

> For the first time in all living time, a living creature understands
> its origin and can undertake to design its future.
> —ROBERT SINSHEIMER, geneticist

> We believe this issue is going to dwarf the pro-life debate within
> a few years. We are on the threshold of mind-bending debates
> about the nature of human life and animal life. We see altering
> life-forms, creating new life-forms as a revolt against God's
> sovereignty, and the attempt by humankind to usurp
> God and be god.
> —RICHARD LAND, executive director,
> Christian Life Commission,
> Southern Baptist Convention

Ever since Darwin, biology has provoked anxiety. Darwinian evolution said human beings were directly related to animals and denied the possibility of a special divine act of creation for each species. In doing so, it inspired a distinctively modern sense of futility. What was the point of life if we were no more than accidental animals—chance products of the blind workings of natural selection?

Many felt that the very fact that such a question could be asked threatened the moral basis of human society. One such was William Jennings Bryan, a lawyer and Presbyterian who had been three times defeated in his attempt, as the Democratic candidate, to gain the presidency of the United States. In the carnage of the First World War trenches, he believed he saw direct evidence of the moral impact of Darwinism: the horror to which the whole of humanity would descend if it became generally accepted that we were indeed descended from monkeys.

Bryan's conviction led directly, in 1925, to the Scopes trial, in which a Tennessee state law against the teaching of Darwinism in schools was challenged by the American Civil Liberties Union through John Scopes, a teacher at Dayton High School. Prosecuted by Bryan, Scopes was found guilty and fined $100—he had, after all, quite clearly broken the law. But it was Bryan who was defeated. Refused permission to call scientists as witnesses, Clarence Darrow, the defense attorney, conducted instead a devastating and now legendary cross-examination of Bryan himself, citing him as an expert on the Bible. Bryan's creationism—the belief in the literal truth of Genesis—crumbled, and the citizens of Tennessee were lampooned as backward and superstitious.

Creationism survived the ridicule, of course. Today it is called "creation science," and there have been renewed attempts to insist that it be given at least equal billing with evolution in the schoolroom. Thinkers like Duane T. Gish have attempted to provide "creation science" with an identity as systematic and persuasive as that of Darwinism.

"The majority in the scientific community and educational circles," Gish has written, "are using the cloak of 'science' to force the teaching of their view of life upon all. The authoritarianism of the medieval church has been replaced by the authoritarianism of rationalistic materialism. Constitutional guarantees are violated and free scientific inquiry is stifled under this blanket of dogmatism. It is time for a change."

In addition, academics like professor John Whitemore at Cedarville College in Ohio are leading a new, young, educated wave of creationists who are forcing scientists to confront an idea they thought had died at the Scopes trial. *Scientific American* magazine noted in its January 1997 issue that almost half the American public believes that God created man within the past ten thousand years, and some surveys have shown that up to 60 percent of Americans believe creationism should be taught in public schools. Creationism, in short, was not defeated by Clarence Darrow, and the traditional faith, to which Bryan appealed, lives on.

But alongside this continued creationist resistance, new ways of expressing the anxiety that perhaps this new biological knowledge is not good for us have emerged.

In the 1960s the first hints of what was happening in biology began to surface in the cultural mainstream. This was the moment at which the popular image of the biologist as a boring old cataloguer and of the physicist as radical, world-changing genius began to be reversed by the news that was coming from the labs—well before recombinant DNA technology made it clear to us all that the manipulation of life at the molecular level was well within our powers. The possibility was there, and, even as early as 1963, the *New York Times* was anticipating these developments:

> The prospect is that in the next few years humanity will understand—and be able to control at least in part—the fabulously intricate mechanism through which each species of living organism transmits its essential properties to the next generation. The danger exists that the scientists will make some of these God-like powers available to us in the next few years well before society—on present evidence—is likely to be even remotely prepared for the ethical and other dilemmas with which we shall be faced.

Note that pessimistic phrase, "on present evidence." In this editorial, the *Times* seemed to be feeling what Bryan felt when he contemplated the First World War: There was something wrong with the modern world. For, although there were no trenches in 1963, there was The Bomb. It was the height of the Cold War, not long after the Cuban Missile Crisis, when that supremely destructive product of modern physics, the hydrogen bomb, seemed to be threatening the human race with annihilation. Bryan saw Darwinism in the trenches; the *Times* saw something more vague, though equally disturbing, in the implacable confrontation of the Cold War. The mad spectacle of a strategic gamble with the apocalypse was the editorial writer's "present evidence" that cast doubt on the ethical status of the human race.

If there was one science above all others that inspired this anxiety, it was physics. Physics, the scientific success story of the first half of this century, had made The Bomb. But it had done more than that; it had created the possibility—the image—of global catastrophe. Even if, as turned out to be the case, the nuclear threat were to recede, the vision of science as global destroyer had been born, and biology was just arriving on the scene to feed that new form of anxiety. As the science historian Charles Weiner has memorably said of this period, "The symbol of the mushroom cloud was becoming intertwined with the symbol of the double helix."

As a result, the awareness of the shift in the balance of power from physics to biology began to filter though to the people.

"For too long," wrote Gordon Rattray Taylor, in his sensationally titled book, *The Biological Time Bomb*, published in 1968, "the public has maintained a false stereotype about biology: it was the science of classification, of botanizing and studying bees. Biologists were dry-as-dust creatures who studied the migration of birds or dissected frogs to see how they tick. In contrast the physicist was seen as much more involved in reality. His tinkerings with sparks and wires produced, in due course, radio and television, not to mention the telephone."

Taylor went on to note the new kind of radical language that was emerging from the biologists. He quotes Joshua Lederberg, then professor of genetics at Stanford University, who said that big changes "in life span and the whole pattern of life are in the offing, providing that the momentum of the existing scientific effort is maintained." Meanwhile, Francis Crick warned: "The development of biology is going to destroy to some extent our traditional grounds for ethical beliefs, and it is not easy to see what to put in their place." Here, startlingly, one of the two atheists who had deciphered DNA was almost exactly echoing the primary fear of William Jennings Bryan.

Biology as physical or moral apocalypse was on its way. On the most obvious level this was another version of the fear of nuclear war. Clearly, biological weapons could be as destructive and nightmarish as nuclear ones; visions of millions of citizens dying of novel plagues designed in the test tube began to compete with the more familiar nuclear scenarios. But biology also offered a completely new kind of threat, subtler and less noisy than the mere detonation of hundreds of nuclear warheads. It arose from a dawning popular awareness of the delicacy of organic systems, an awareness that was later to find full expression in the environmental movement. In the late 1950s, for example, Rachel Carson's book *Silent Spring* was published, exposing the way the use of chemicals such as the pesticide DDT could have huge, unexpected side effects. While scientists had insisted that the quantities of DDT required to kill undesirable insects were too small to do further damage, these chemical traces could, through the workings of the food chain, accumulate in the bodies of birds, fish, and humans. Thus the biological system on which we depended was vulnerable to attack, not only from missiles but also from molecules. Biology, like physics, endorsed the new popular awareness that the very smallest things— molecules and atoms—could have the most devastating effects.

Movies like *The Andromeda Strain* in 1970 started a popular tradition of contemporary biological anxiety that persists today in films like *Outbreak* and *Jurassic Park*. All are based on the terrible possibility that, in tinkering with the fundamentals of life itself, we may be destabilizing the whole intricately balanced system that keeps us alive. Nature will not be mocked, runs the underlying moral of all these fictions, and she will take her revenge.

But from the beginning this was not just a case of profitable paranoia-generation by the controllers of popular culture. For the scientists themselves were also suffering anxiety attacks. "When man becomes capable of instructing his own cells," wrote the biologist Marshall Nirenberg in 1968, "he must refrain from doing so until he has sufficient wisdom to use this knowledge for the benefit of mankind. I state this problem well in advance of the need to resolve it because decisions concerning the application of this knowledge must ultimately be made by society and only an informed society can make such decisions wisely."

And another biologist, Leon Kass, wrote of "a technology whose consequences will probably dwarf those which resulted from the development of the atomic bomb." Such words were especially provocative in the midst of the political radicalism of the sixties, when not trusting your parents, the government, scientists, or even people in general was the standard posture for the most politicized of the younger generation. Biology became one more wicked plot of the old and the powerful.

In 1969, in the midst of this atmosphere of dissent, a team from Harvard Medical School identified a gene complex known as the *lac operon* in the bacterium *Escherichia coli*, a normal inhabitant of the human gut. This was an important scientific moment since the *lac operon* was the first device to give an indication of how gene control mechanisms work. But it was also an important political moment. In announcing this new development, the team leader, Jonathan Beckwith, did not speak of a scientific triumph, but rather of a scientific threat: "The more we think about it," he told the *New York Times*, "the more we realize that it could be used to purify genes in higher organisms. The steps do not exist now, but it is not inconceivable that within not too long, it could be used, and it becomes more and more frightening—especially when we see work in biology used by our government in Vietnam and in devising chemical and biological weapons."

That a scientist should say such things at what must have been one of the great moments of his career is an indication of the anxious climate of the time and of the way biologists themselves felt a degree of awe and uncertainty at the progress of their science. Beckwith's evocation of Vietnam was important, because that was a war which had engendered deep mistrust between government and intellectuals. It was a war in which chemical defoliants were used—a foretaste of the biological apocalypse. The war was a turning point not just for American politics and society, but for American technology as well.

These then-inchoate but acute anxieties did result, in the late sixties and early seventies, in a number of ethical initiatives designed to study the impact of the new biology. But the techniques of recombinant DNA that emerged in the early seventies transformed the picture; one moment there was quiet, mounting concern; the next there was real alarm. For recombinant DNA made it clear that the future, forecast by the *New York Times*, had arrived. What had been science became technology. We could manipulate life at the molecular level. We could create biological novelty. Anything might be possible.

But, perhaps oddly, the primary anxiety among scientists in the seventies was not what we would do with this knowledge, but rather what it would do to us. New organisms, such as viruses and bacteria, might be created whose effects would be unknown. Maybe they would cause cancer or some novel infectious disease which we would be powerless to resist. Or maybe they would attack animals, plants, or even other microorganisms and destroy the ecological balance of the world, making it unfit for life.

A hesitant moratorium on recombinant DNA research in 1974 ran immediately into pressure for commercial exploitation of the new technologies. Two university biologists applied for a patent on the technique of recombinant DNA, thereby forcing the issue. It was clear that the scientists had to show they were responsibly assessing the risks involved, so, in January 1975, they gathered at Asilomar, California. The opening remarks by Nobel Prize–winning geneticist David Baltimore summarized the portentous theme of the meeting: life would never be the same again. "Although I think it's obvious that this technology is possibly the most potent potential technology in biological warfare," he said, "this meeting is not designed to deal with that question. The issue that does bring us here is that a new technique of molecular biology appears to have allowed us to outdo the standard events of evolution by making combinations of genes which could be immediate natural history."

Baltimore's awkward phrasing betrays an effort to avoid being too sensational. But behind "outdo the standard events of evolution" and "immediate natural history" lurks the simple and staggering admission: we are taking control of life. There is also something odd about Baltimore's statement of the purpose of the conference. After all, why *weren't* they discussing biological warfare? At a time when a nuclear war was still a real possibility—when, in other words, a new and radical technology was threatening to destroy mankind—it would seem only natural that the first thing on the scientists' minds when they got together to discuss the dangers associated with another new and radical technology (recombinant DNA) would be the potential for using it to construct biological weapons. This would seem to be the most pressing risk factor, yet the assumption at Asilomar was that the risk that most needed to be considered was the accidental, or nonmalicious, effect of the technology. The point was, of course, that this new technology might have a life of its own that would subvert all human decisions. War was a choice, cancer was not.

A statement had been issued about biological warfare prior to the Asilomar meeting by scientists of the Plasmid Working Group. They were clear about the most likely danger. "We believe," ran the statement, "that perhaps the greatest potential for biohazards involving alteration of microorganisms relates to possible military applications. We believe strongly that construction of genetically altered microorganisms for any military purpose should be expressly prohibited by international treaty, and we urge that such prohibition be agreed upon as expeditiously as possible."

Yet, for whatever reason, Asilomar was not to be about the hazards of war but about the hazards of peace. The problem was that these hazards were completely unquantifiable. As all the scientists at Asilomar knew perfectly well, the process of taking control of life was in its very earliest stages. Recombinant DNA technology had revealed a vast landscape of potential, but the landscape was dark. It might contain rogue viruses that would infect lab assistants with cancer, or bacteria that would destroy the U.S. wheat harvest, or devastating new global plagues that would fling the planet back to the Dark Ages. Just as anything good suddenly seemed possible, anything bad was also in the cards.

The key figure at Asilomar was Stanford geneticist Paul Berg. Since 1968 he had been working with a monkey tumor virus called SV40. The virus had been first isolated from the kidney of a rhesus monkey in 1960: This caused a panic, as this particular monkey organ was used for the production of polio vaccine, which meant that millions of people may have been infected with SV40. Because of the rate of development of cancers, nobody would know what damage had been done for years.

Yet SV40 remained a valuable tool. It seemed to be able to shift genetic material about.with remarkable flexibility. Berg wanted to use it as a carrier to move genes from one cell to another—potentially a way of developing medical gene therapy. But, amidst this shuffling of genes, who could know what might emerge? Might the people in the labs be infected by either SV40 or some novel hybrid virus? It was Berg's concern about these questions that led to the Asilomar conference. And it was Berg who was to give his name to the primary document that eventually emerged from the discussions—the Berg Letter. This letter called on all scientists to observe a moratorium on certain types of recombinant DNA experiments until an international conference could be held on the potential risks. In the eyes of the lay world, the scientists had effectively and officially endorsed biological anxiety.

Asilomar and the Berg Letter are often now spoken of as turning points—landmark events that demonstrate conclusively the farsighted responsibility of the scientific community. They happened in the shadow of The Bomb, a doomsday machine whose existence was the sole responsibility of science.

"It was quite unprecedented in the whole history of science for a group of scientists to call for a halt to their work, and trust to the force of consensus to ensure that colleagues in other countries did not cheat," wrote the philosopher Jerome Ravetz. "If Leo Szilard had been successful in his efforts to get such a moratorium among the much smaller group of atomic scientists in 1938, the subsequent history of the world would have been much simpler and safer."

Again there is that link between nuclear and biological dangers. Ravetz says that an Asilomar-type scientific "strike" in the late thirties might have prevented the Cold War.

Erwin Chargaff—who had paved the way for the analysis of the structure of DNA and who identified the quantities of the chemical bases in samples of DNA that led directly to the unraveling of the molecular structure by Watson and Crick—mistrusted the reductionism that claimed that the molecule was "the secret of life," and, as a result, was skeptical of many of the insights leading from that discovery. He summarized the primary fear: "Have we the right to counteract, irreversibly, the evolutionary wisdom of millions of years, in order to satisfy the ambition and curiosity of a few scientists?" And he also warned: "I am one of the few people old enough to remember that the extermination camps in Nazi Germany began as an experiment in genetics."

Chargaff's remarks stand as representatives of the central biological fear of the seventies—the fear that this knowledge, like the physics that produced The Bomb, was too much for us to handle.

Asilomar was an extraordinary moment in the history of science; there certainly was something unprecedented about the conference and the Berg Letter. The scientists did act to try to prevent and to warn of unknown dangers, and they did expect other scientists to cooperate. It was an idealistic statement of the supranational responsibilities of science. It was also a moment at which the deepest foundations of scientific confidence seemed to be shaken. The usual argument advanced by scientists when they are attacked about the results of their work is that knowledge is neutral, that it is the application of that knowledge which involves moral, social, and political issues. So Einstein cannot be blamed for The Bomb; he merely did the fundamental physics. If his physics was then used to make a bomb, it was because human society is corrupt and fallen, not because there is anything inherently wrong with science. Between the scientist and the application of his work stands society.

But Chargaff explicitly—and the Asilomar scientists implicitly—were suggesting that the new biology might threaten this convenient argument. Perhaps this knowledge was *not* neutral. Perhaps merely pursuing this knowledge, without ever applying it, was dangerous. Simply researching recombinant DNA meant making novel genetic structures, one of which could accidentally escape and cause untold damage.

So recombinant DNA represented the crossing of a barrier. Science, after almost 400 years, had finally found a way of manipulating the basic elements of life. Living matter was like everything else, and, like everything else, we could change it. Could such knowledge really be classified as neutral?

The nervousness exhibited by the scientific community at this time was expressed, variously, as a fear of military applications, of rogue viruses escaping from the lab, of the destabilization of 3.5 billion years of evolution, and so on. In other words, it was expressed as a series of specific fears and possibilities. But beyond this, I suspect there was a deeper, more indefinable anxiety: Recombinant DNA was a step into the unknown, but not the unknown of space or matter; it was the unknown of life. And however successful scientists had been in eradicating the last traces of vitalism (the belief that there was something special about living matter) from their imaginations, something lingered—some sense that maybe this was not just physics and chemistry as usual. Frankenstein, Mary Shelley's man-made monster, seemed to persist as the clearest possible warning that combining technology and life processes could be catastrophic.

In the wake of Asilomar, public anxiety about the new biology rose. At Harvard in 1976 a bitter battle was fought over the siting of a new genetic engineering laboratory. This produced a wave of local government concern across America about the possible dangers of recombinant DNA technology. In 1977 a symposium at the National Academy of Sciences in Washington, D.C., was disrupted by protesters from the Coalition for Responsible Genetic Research, who carried banners saying "We Shall Not Be Cloned" and one that quoted Adolf Hitler: "We Will Create the Perfect Race." Under this kind of pressure, government seemed to be heading for strict controls on the work of scientists.

But if government, the people, and even scientists did feel a chill, metaphysical wind in the 1970s, it did not last long. Their anxious phase is now seen more as a philosophical than a practical matter, for today the fears expressed at Asilomar have been largely forgotten within the scientific community. The unquantifiable risks are now regarded as being unquantifiably small. "There is no evidence," said James Watson, "that anybody ever got sick from a recombinant DNA experiment."

Take, for example, the release of a genetically modified organism into the environment. The "wild" organisms it would encounter are the product of millions of years of evolution and are, as a result, robust and highly adapted, their phenotypes fine-tuned over thousands, millions of generations. The chances of an artificially or accidentally modified organism succeeding in this environment are minimal. Even if a scientist deliberately designed what he thought would be a successful genetic modification, he would be unlikely to get it right. So subtle and efficient are the effects of evolution that apparent improvements are hard to find; in any case, they might create unexpected problems for the organism.

This is not, of course, to say that we cannot successfully change nature within tightly managed situations. In agriculture, we have been doing it for thousands of years, and, since the seventies, we have been doing it genetically. Genetically engineered food is now common on supermarket shelves. And it is not to say that organisms cannot prosper in wholly alien environments. In Britain, an imported plant— the rhododendron—has spread uncontrollably at the expense of local flora. Ships' cats belonging to the first white explorers and settlers in Australia devastated local small animals, and donkeys from the California Gold Rush drove bighorn sheep out of the Rockies. If such things can happen at the level of the large organism, they can, in theory, happen at the level of the small.

But the specific fears of Asilomar have faded because they now seem overstated, from both a cultural as well as a scientific perspective. The anxiety felt in the fifties, sixties, and seventies about the fragility of the natural environment was linked to the apocalyptic climate of the Cold

GOD, THE BOMB, AND THE DOUBLE HELIX · 37

War; that, in turn, stimulated a general mistrust of science and technology. The Cold War is over and some of that mistrust has faded. The triumph of the West over communism has, to some extent, validated our belief in our technological culture. It is less common today to be so anxious about science and technology; rather, primarily because of the success of information technology, people see them as providing exciting new ways of living their lives. Genetics has benefited from this change. Its public image is now determined largely by the steady flow of stories about genetic discoveries that offer hope of medical treatment or horoscope-like insights into human behavior. These stories all tend to fall into the science correspondent's familiar genre of "good/weird news from the labs." It is almost certainly true that the average citizen does not feel queasy or alarmed when he reads of the discovery of the gene that predisposes one to muscular dystrophy or cystic fibrosis. He has become accustomed to the basic fact that we are now able to read the script of life.

But, that said, anxiety persists. Most familiarly it is expressed as an "ethical" concern. When Dolly, the cloned sheep, first appeared, for example, President Bill Clinton's first move was to call for an ethical report. He would not have done the same if the Intel Corporation had announced the production of a radical new computer chip. The science of life demands a different response, an acknowledgment of anxiety. Both in Britain and the U.S., ethical committees are constantly deliberating over the morality of the latest biological initiatives.

This is a strange spectacle, for, although the word "ethics" is repeatedly employed, its meaning is far from clear when used by public figures. Do they mean a single moral system that could be applied to us all? Or do they mean a specific system that is intended for use within one area of expertise or policy? The distinction is important, because a local ethical system does not demand a general statement, only an expression of its utility within a particular discipline. A general ethical system, in contrast, implies a moral basis that is common to us all. Deciding on the ethical implications of Dolly affects the whole of society. And society, in liberal democratic terms, is not one system but many. Indeed, the whole point of a liberal democracy is that it has no ethical positions beyond the minimum needed to sustain itself. The success of

the West is based on the fact that, unlike communist, fascist, or theo-
cratic societies, it does not impose a single moral vision of the world
upon its inhabitants. In this way we can hope that people of widely dif-
fering beliefs can live peacefully together.

Much of the time, of course, we pretend otherwise; we pretend we
are more unified than we actually are. Political rhetoric needs at least
the illusion of moral visions. But when politicians are faced with the
strong moral feelings of various special-interest groups—say, the prolifers and the prochoicers of the abortion issue—they find these illusions
exposed. The vague moral vision of the political speech collapses when
confronted with such a dilemma. To adopt either ethical stance would
alienate too many people. And, of course, the passionate advocates of
both positions find the resulting neutrality of the state almost as repellent as the ideas of their opponents.

As a result, when politicians ask for ethical advice on some specific issue, it is not at all clear what they mean, what they want, or what they
intend to do with this advice. The history of ethical thought is long and
complex, but it has produced no system to which everybody in a secular,
democratic state is likely to subscribe. There are only opposing forces, as
in the abortion debate. The process of counseling politicians tends,
therefore, to follow the same pattern. A technological advance—say,
human cloning—is considered. Those against and those for are heard. A
committee in some way takes the temperature of public feeling on the
issue, perhaps by reading newspapers or by taking a poll, and then produces a report. The report will be a compromise between various levels
of unease—from extreme revulsion to mild concern—and the demands
of business and technology. It will, in other words, be a simple balancing of forces. Ethics, in any recognizable sense of the word, does not
come into it. Perhaps that is the only way there can be any public discussion of morality, but, if so, we should be aware that we are not really
discussing morals, we are talking politics.

I make this point at some length because I think it provides a necessary background for a discussion of the way resistance to genetic innovation now finds expression. In a world where ethics is replaced by a
balancing of forces, then the best way to advance an ethical position is
with the maximum force. This is what happens in the abortion debate,
where some prolifers have resorted to killing doctors, and it is, to some

extent, what is happening in the genetics debate. Indeed, some say genetics is going to be an even bigger battleground than abortion.

In May 1995, Dr. Richard Land, executive director of the Christian Life Commission of the Southern Baptist Convention, announced the formation of a coalition "in vigorous opposition to the patenting of human genes and genetically engineered animals." A large number of religious leaders had signed a statement that said: "We, the undersigned religious leaders, oppose the patenting of human and animal life forms. We are disturbed by the U.S. Patent Office's recent decision to patent human body parts and several genetically engineered animals. We believe that humans and animals are creations of God not humans, and as such should not be patented as human inventions."

Land's statement made it clear that they were not opposing genetics on principle. He welcomed work "aimed at the treatment and cure of hereditary diseases" and insisted he had been "generally supportive of the advance of genetic science." But patenting was a step too far; it implied that animal and human life could be owned, whereas it was, in fact, "pre-owned" by God. Patenting implied a degree of control over nature that seemed to move the whole living realm into a legal, commercial context.

For the same reason that patenting was unacceptable, transgenic experiments, in which animal genetic material is introduced into human material, "devalues human life and, in our view, represents a form of genetic bestiality." In addition, the "ethical utilitarianism" that found such experiments acceptable was threatening to create a "brave new world" or programmed "chimeras or parahumans," effectively slaves. And, finally, patenting "commodifies" people: "Marketing human life is a form of genetic slavery." We can be rescued from all this only by "an ethic of the sanctity of human life that will protect human persons against the tyranny of the technological imperative."

This is a very broad statement which amounts to a significant refinement of William Jennings Bryan's resistance to Darwinism. Clearly, the basis of this new form of religious resistance has much in common with Bryan's position; like Bryan, Land and the signatories to the declaration feel that there is something about life that lies beyond mere science. Indeed, Land's statement says, "Our candid presupposition is that human beings, and even animals, are more than their sum of DNA." This is somewhat incoherent; if people are more than their DNA, then manipulating that chemical need not necessarily be so bad. But clearly the fear is that science threatens to reduce people to their genetic material and, in doing so, to destroy their God-given moral autonomy.

At the time of the Scopes trial few people had heard of DNA, and none could have been sure that it was the genetic messenger. Bryan had a generalized fear of Darwinism, not a specific fear of the power of molecular biology. But our new genetic knowledge demands a new definition of religious resistance. Instead of standing on the difficult ground of the literal truth of the Bible, Land and these other new religious dissidents rest their case on the more general issue of the "sanctity of human life." They are not, in other words, saying the science is wrong, but that it is all too right. This rightness, however, does not give it carte blanche to interfere with the divinely ordered processes of life. Unless life is placed beyond the interventions of science, there can be no limit to the malign transformations of moral order that it can impose. Land himself bases this belief on a "biblical world view," but, he insists, it is a belief that it is held by many who do not share that view. The backing of faith is not strictly necessary. The atheist can also draw a moral line around life.

This form of the resistance to biology has proved popular. Indeed, the most vociferous and effective resistance to genetics has come from outside any religious establishment. Jeremy Rifkin's Washington, D.C.-based Foundation on Economic Trends has pursued a consistent war against genetics and biotechnology that has successfully whipped up public concern. Rifkin is a brilliant campaigner who insists that he is not trying to stop progress, but rather to make sure the public understands all the implications of the new technologies before they are approved by scientists and legislators. In 1983 he successfully, though temporarily, brought to a halt trials of a genetically modified bacterium that was intended to prevent ice damage to trees in California. And he has led high-profile campaigns against genetically engineered food that have led to widespread public suspicion that we are not being told of all the potential dangers.

Rifkin's talents and his broad resistance to almost any biological intervention have resulted in serious loathing from the scientific community. *The Genetic Revolution*, a book of essays on the future of genetics, begins with a specific attack on Rifkin's tactics by Bernard D. Davis in his introduction:

> While the contributors to this book disagreed somewhat about the assessment of risks, they did agree on one feature of public debate: the need to combat the influence of those critics, such as Jeremy Rifkin, who are ideologically opposed to all genetic engineering (though in specific cases they often present their objections in more moderate terms). Their appeal to public uneasiness about technology in general, and about the unfamiliar microbial world in particular, has made them highly newsworthy; and they have had a good deal of political influence, especially the Green political party in Europe. Unfortunately, scientists are generally unskilled at public relations and find it difficult to deal with professional activists. The public is therefore often exposed to a distorted picture that is only slowly corrected.

The scientists' problem is that, though it might not be possible to whip up antigenetic resistance on the basis of the literal truth of the Bible or even on the basis of Land's more subtle religious position, it is certainly possible to appeal to a general environmental anxiety. Everybody now worries about the future of the planet. We have, say the environmentalists, gone too far in our attempt to control nature. In such a context it becomes relatively easy to suggest that the advancement of genetics might simply be our latest act of going too far.

Even the Prince of Wales is able to express worries about genetics. In a speech in December 1995, he spoke of being "profoundly apprehensive" about the effects of genetically engineered crops and of the "confidence bordering on arrogance" of the way biotechnology companies were promoting their products: "Is it fair to ask whether anyone has given any real thought to what to do if . . . herbicide-resistant plants spread to become weeds in fields of other crops? Won't the inevitable answer be new and stronger herbicides, leading to further reductions in biodiversity?"

The prince did not, in this speech, stray into the realm of human genetics, and his reasoning was primarily based on the practical point that we might be making mistakes that have disastrous consequences. But, in light of his attempts in other speeches to encourage a spiritual view of life, it is clear that the inspiration for these remarks is not purely practical. He is attempting to evoke something like the idea of the sacredness of life to which Land directly refers. The prince, significantly, speaks of the way some of these developments "send a cold shiver down the spine." A cold shiver suggests something more than a merely practical unease.

So now the moral position on genetics, like the scientific one, seems poised on the edge of an uncertain future. The strength and depth of support for the hard-line resistance of either Rifkin or Land, or for the more vague anxieties of Prince Charles, are unclear. Much depends on the political climate. Republicans, for example, may be dominated either by the demands of big business—including, of course, the biotechnology companies—or by the fears of the Christian right about usurping the position of God. The Democrats, on the other hand, may be on the side of genetics as a progressive, technological path toward improvement of the human race, or they may oppose it as one more environmental abuse.

The problem is that neither God nor the environment is a match for the strong technological determinism of our day. They both lack the huge, persuasive promise of a science that seems to offer both a cure for cancer and an end to human hunger and poverty. Equally, more generalized resistance to biological intervention, like the "maintaining the dignity of the individual" clause that appears in French biotechnology laws, can be open to many different interpretations. Is, for example, the dignity of the individual better maintained by leaving his genome alone or by pursuing research that might relieve his suffering or that of his children?

Evidently many people want to draw lines, but, equally evidently, they are uncertain about how or where to draw them. We can believe in one specific orthodoxy, as in the appeal to the Bible, or we can be generally concerned, as is the Prince of Wales with his "cold chill down the spine." Neither position is universally persuasive. In the resulting uncertainty, the hard certainties of science tend to appear as the easier option.

Perhaps, people reason, we have learned our lessons, from the trenches, The Bomb, or environmental destruction, and now we have grown up enough—now we know enough—to handle technology responsibly. There may be nothing to fear from our mastery of life.

chapter 3

Eugenics 1: The Right to Be Unhappy
Eugenics 1: The Right to Be Unhappy

"In fact," said Mustapha Mond, "you're
claiming the right to be unhappy."
—ALDOUS HUXLEY, *Brave New World*

There is one genetic anxiety which I so far have not mentioned. This is neither a vague unease nor an unquantifiable laboratory risk nor a religious conviction that we are trespassing on God's territory. Rather, it is based on a sober assessment of the evidence of human history. We may not know for certain whether our science angers God or risks new plagues, but we can be absolutely sure that genetics has been used in the past to justify the most horrific crimes—mass murder, mass sterilization, and racial and cultural hatred. Eugenics—the attempt to improve the human species by altering its reproductive habits—is an old idea. In Athens, Plato advocated it as a necessary aspect of the ideal society: The best should mate with the best and the worst prevented from reproducing. It was the rational thing to do.

In the fifteenth and sixteenth centuries, the Spanish Inquisition—the brutal defender of Catholic orthodoxy—had a very precise, eugenic definition of the "best." It spoke of *limpieza de sangre*—the purity of the blood. This purity, it was believed, was the ultimate biological distinction between Christians and Jews: The blood of the Jews was impure and should be prevented, at all costs, from tainting good Christian stock by cross-breeding.

From the point of view of an intense, prescientific, unquestioning and unquestionable faith, there is something appallingly reasonable about this: Christianity was the absolute, revealed truth, yet here were these people, the Jews, who refused to accept this revelation. They must be different, it was thought. Something within them forced them to deny the truth. And they were not just different internally—they also looked different. They were a separate race. They were *made* differently from Christians.

The conclusion was obvious. Jews did not accept the Christian revelation because of something in their biology. It was not a choice; it was forced upon them by their physical substance. And, since the Christian revelation was true, then that biological fact, whatever it was, could not be just a benign expression of human variety; it was a flaw. Knowing nothing of the mechanism of heredity, the Inquisition fell back on an ancient idea—that this fault was to be found in the blood. The blood of the Jews was impure, while that of the Christians was pure and in harmony with the truth of the universe.

Such a conclusion appears to be in accord with the facts of faith, race, and Christian justice as interpreted by the Inquisition. Of course, it is also vicious, racist, and murderous. From a modern, enlightened perspective, the idea of the impurity of the Jews is repellent, evoking visions of Auschwitz and the Stalinist pogroms. And, scientifically, we know it to be absurd. Even if we detect observable differences in the DNA of the Gentile and the Jew, there is no basis for ascribing purity to one and impurity to the other. At the molecular level we do not find such clear moral signposts. So, even if we want to be anti-Semitic, we have found in the ultimate constituents of life no evidence for our bigoted desire—no justification for racial eugenics.

This may be a comforting thought, but it should not be too comforting. For although geneticists do not seek out Jewishness in our blood, they do seek out other things. Where those priests saw faith or heresy in the blood, our scientists now aspire to see schizophrenia, alcoholism, depression, or risk-taking behavior in the genes. The idea is, in essence, the same. Some aspect of our biology determines some aspects of our behavior, personality, or even opinions. What we are is what we appear to be: a physical entity. And what we do and think springs directly from what we are.

Finding such a parallel between our enlightened selves and the barbarity of the Spanish Inquisition may seem disturbing or absurd. Those who think it disturbing will argue something like this: Precisely because a belief in fundamental biological differences has led to such horrors in the past, and precisely because it is obvious that such knowledge was deliberately rigged to provide a spurious objective basis for bigotry, we should be very, very cautious about using biological differences to explain behavior, personality, or even disease. The history of biological justifications is a bloody one, far too bloody for us ever to contemplate taking such risks again.

Those who find the parallel absurd will argue something like this: The Inquisition was ignorant of the true facts of biology, and its motives were disgusting. We know the important biological facts and our motives are sound: We wish to identify the genes that cause schizophrenia in order to cure that frightful condition. This argument has been put forward by James Watson. I think it is naive and so does the historian Mitchell G. Ash, who comments, "James Watson has articulated a widespread view that incomplete knowledge was abused in a truly cavalier way by the Nazis. But this implies, first, that more complete knowledge would somehow be less likely to be abused—yet the naivete of that claim is clear as soon as it is stated. The conventional view implies, second, that what knowledge there was was abused only after it had been gathered—that is, that 'good' science could not have been done under an evil regime."

I am with Ash. The Nazis had more complete biological knowledge than Darwin, but it was they, not Darwin, who murdered the Jews.

You may find either of these arguments convincing. Or you may not think they are so very far apart. After all, those who hold the second belief that we are, indeed, very different from the Spanish Inquisition can, at the same time, accept that we should be very cautious about how we handle genetics in the light of our knowledge of the past. But this would be a mistake. Although, logically, these two positions can be made to agree, they actually represent profoundly different attitudes. The second is progressive and makes optimistic assumptions about the moral improvement of the human species; the first assumes that human beings do not change so very much, that all knowledge can be abused, and that we should be constantly aware of history. How we feel about eugenics, what we think eugenics is, will depend on where we stand on that divide.

Let me go back to this question of the ultimate biological basis for our behavior. Schizophrenia or alcoholism may be in the genes; we do not yet know for sure. But the research community, encouraged by the vast sums of cash invested by drugs companies anxious to find a cure for both these conditions, believe that both have large and probably decisive genetic components. (It is worth noting here that one of the originators of the idea that schizophrenia was genetic was the German Ernst Rudin, who served on a panel, headed by Heinrich Himmler, that helped draw up the Nazis' sterilization law of 1933.) While the priests of the Inquisition took the difference between Jewish and Christian blood to be a matter of faith interpreted by reason (there *was* an ulti- mate difference, *therefore* it must be in the biology), we take the differ- ence between schizophrenic and nonschizophrenic DNA to be a matter of observation and experiment. Our method seems to work better; it gets results. It has, oddly enough, even proved the Spanish Inquisition almost right. There *is* a biological difference between Jews and Chris- tians; it is small and it may not be very consistent, but it is there if you want it.

So we cannot easily distance ourselves from those Spanish priests simply by insisting they were wrong and we are right. The objective content of their idea that there are biological differences between peo- ples is, in modern terms, correct. Equally, we seem to agree with their view that these differences result in fundamental variations. They said there was something in the blood of the Jews that prevented them from seeing the truth of Christianity. We say there is something in the genes of schizophrenics that leads to their aberrant behavior. It is no good saying, Ah, but schizophrenia is an illness and Jewishness is not, be- cause, to the Inquisition, Jewishness was at least a sickness if not some- thing far worse.

Some go even further in insisting that fundamental spiritual variations can be explained in terms of genes. One scientist has expressed the view that there is a "metaphysical gene"—some stretch of DNA that makes some of us prone to religious belief and others immune to it.

Both the modern geneticist and the Inquisitorial priest are driven by the conviction that the aberration they detect in the person in front of them must be rooted in some material defect.

Such a view may seem obvious in the case of, say, blindness—the scientist may examine the eyes and explain why they do not see. But in the case of non-Christian religion or a mental illness, the situation is less obvious. Priests or scientists may *feel* with greater or lesser amounts of evidence that there is some material basis for this aberration. But, for the moment, whether one accepts this feeling or not remains a matter of choice. The metaphysical gene may or may not exist; the genes for schizophrenia are assumed to exist. Absolute proof is not available.

It is clear, however, that this is not a choice to be taken lightly. For the priests to decide you had Jewish blood might mean your death. Certainly many millions, throughout history, have died because some people concluded they were fundamentally, irredeemably inferior. It was, after all, the Jews once again who suffered because the Nazis decided they were biologically a lesser race. And the book that Adolf Hitler read while in prison after his first, abortive attempt to gain power was the standard genetics text of the day, *The Principles of Human Heredity and Race Hygiene* by Eugene Fischer. Fischer was the director of the Berlin Institute for Anthropology, Human Heredity and Eugenics, where one of his assistants was Joseph Mengele, that foremost exponent of the science of the concentration camps. Fischer's *Principles* may have been the standard text, but it was, we would now say, wrong. Nevertheless, it was used to justify the most savage genetic experiment in human history: the Final Solution—the extermination of the "genetically inferior" Jews.

The point is that once people decide you are a lesser creature for whatever reason, either superstitious or scientific, there appears to be no limit to what cruelty they may inflict on you. And they are likely to inflict that cruelty feeling fully justified, because it is but a small step from believing another human being is inferior to believing that he is bad, dangerous, or threatening to "superior" beings. Indeed, some may generalize the point even further and insist that all "inferior" beings are dangerous because they threaten the life or health of the entire human race. They may then advocate sterilization, restrictions on marriage, or even murder to prevent the outcast's assault on the integrity of the species.

All of the foregoing is intended to give some idea of the depth, complexity, and danger of the issue of eugenics. Eugenics means, literally, good breeding. The names Eugene and Eugenia mean "well-born"; they celebrate the parents' belief that their offspring are especially blessed. Put in such a way, the term seems innocent enough, parental love, for either environmental or genetic reasons, clearly establishes a reasonable basis for claiming that a child is "well-born."

But when we step out of the charmed circle of familial love into the outside world, the idea of being "well-born" immediately takes on all sorts of troublesome connotations. We may approve the specific love of the parents for one particular child, but, in the public realm, to make qualitative distinctions between people is difficult and dangerous.

We always have made such distinctions, of course. Throughout history people have been driven to fight and die on the basis of the belief that some other tribe, creed, race, or nation is morally inferior to their own. Even when the real justification for warfare is economic or strategic, the rhetoric of war is invariably moral. In the Persian Gulf War of 1991, the Allies did not free Kuwait from Iraq—nor did Iraq invade Kuwait—with cries of "Oil!" but rather with cries of "freedom," or, in the case of the Iraqi invasion, with cries of "United Islam!" In addition, most societies have been based on the conviction that there are natural differences between people. Any aristocratic system presupposes that there is something "in the blood" that distinguishes aristocrats from the rest of the population. Slavery was justified on the basis that some people were born to be slaves. The preservation of an aristocratic line by the careful selection of marriage partners is the purest eugenics. And all the romantic tales of rebellion against such practices—so common in the nineteenth-century novel—show the conflict between the eugenic, aristocratic ideal of continuity and the rising, democratic ideal of equality. Indeed, some of the problems of the British royal family have been ascribed to the fact that Prince Charles and Prince Andrew both broke away from this convention by marrying Diana Spencer and Sarah Ferguson respectively, and, by behaving uneugenically, brought on themselves the woes that followed. Who knows what cruel, clairvoyant irony led Sarah and Andrew to christen one of their daughters Eugenie?

Two important general points arise in any consideration of the fundamental human issues raised by eugenics. The first is philosophical, the second psychological.

The philosophical point to be borne in mind centers on the degree of value we attach to the human individual. In aristocratic or slave-based societies, it is evident that some individuals are seen by others as intrinsically less valuable. The peasant is less significant than the lord, the slave less worth saving than his master. Without making such a distinction either explicitly or implicitly, the societies could not continue to function. However, Christianity undermined this assumption by insisting that all people had immortal souls and that these would be judged by God irrespective of their social status. There was a justice of heaven which transcended the social arrangements of Earth. This was potentially subversive of almost any conceivable social order other than the purest democracy or total anarchy. If the heavenly order was to be reflected on Earth, then no man could rightfully claim any natural superiority over any other—indeed, *natural* superiority was the one thing he could not have. And even if he claimed merely cultural or educational superiority, this claim would seem thin and vacuous next to the supposedly inferior man's claim to be equal in the sight of God.

But, in practice, certain social orders managed to adapt Christianity to their own ends. For example, the idea of heavenly rather than earthly reward was used by aristocrats and other elites to justify inequality and suffering, saying that all would be put right in the afterlife. To complain of one's place in the world demonstrated insufficient faith in the ultimate justice of heaven.

The era of modern philosophy, from roughly 1600 onward, has been one in which the Christian conviction of the equal value of all people has been tested in a new way. The eighteenth century, known generally as the Enlightenment, was one in which science made a devastating assault on the old religious order. The heavens were found not to be arranged in the way the Catholic Church said they were. The physics of Newton and Galileo was completely different from the physics of Aristotle and Aquinas. The Christian world moved from a vision in which every aspect of material reality advertised the living presence of God to one in which He seemed to be almost an irrelevance; He may still have been there, but He was not essential to the process.

The scale of the moral change involved at this time cannot be over-stated. The medieval mind might apprehend the stars or a flower and see clear evidence of the hand of God in their pattern and design. But the Enlightenment mind saw only the remote possibility of such a pattern and design. Shocked by the invasions of science, the cautious Enlightenment man could only say that certain things remained inac-cessible to human reasoning. Possibly such things were evidence of God's existence, but equally possibly they would become accessible to the mind of man once science had made further advances. The point was that we could no longer be sure that what we saw was beyond rea-son or that it required God's guiding intelligence.

Thus the world ceased to tell people anything of substance about how to behave. Indeed, nature began to seem positively immoral. In the nineteenth century this was a particularly painful insight. Stephen Jay Gould has written of the discovery of the behavior of the ichneu-mon wasp. The female ichneumon injects her eggs into, usually, a caterpillar. The egg becomes a larva which then eats the caterpillar in a specific pattern designed to keep the creature alive for as long as the larva requires. Such behavior would, in human terms, be seen as intol-erably savage, brutally calculated to inflict the maximum suffering. But logically, these words are meaningless since we can hardly regard the wasp larva as a moral agent. And yet, as Gould points out, we can scarcely describe the wasp's behavior without turning it into a story with moral and mythic overtones. We look at the world through moral eyes and, in doing so, we are invariably doomed at least to disappoint-ment and, more often, to horror.

So morality was not present in the facts of the world. Nature could not tell us how to behave. "Ought," the moral injunction, could not be derived from "is," the objective truth of the world.

The greatest of all the Enlightenment philosophers, Immanuel Kant, took this revolution of the human imagination and used it to forge an entirely new metaphysical system in which the facts of the world were not merely morally neutral but also radically unreliable. They were the products of our senses—no more than phenomena arising from the way we happened to see things, not the things themselves. But, Kant said, every human has a built-in moral direction, a sense of rightness and purpose. Every person is, therefore, an entity of total moral significance and responsibility; each must make his moral choices as if they were instantly to become the moral law for everyone. The moral issue is always: What if everybody did this?

If every individual was an end in himself, if he could not be regarded as a means to an end—as a tool or pathway for others—then the moral importance of the individual was absolute. This idea remains at the center of all moral debate. You can be for or against Kant's decisive distillation of the moral issues of the modern world, but you cannot possibly ignore it. The central issue of eugenics is that it raises doubts about the individual as moral absolute and threatens to overturn the moral basis of Enlightenment thought.

The psychological point to be considered in discussing eugenics involves the way we tend to regard other people. However unprejudiced and enlightened we may think we are, it is impossible to live one's life without making judgments about our fellow human beings. These judgments may be right or wrong, but they will almost certainly be on the basis of inadequate evidence. You may idly classify as "an idiot" a driver who passes recklessly, or you may dismiss as "inconsiderate" somebody who plays music too loud late at night. In neither case do you really have enough information to arrive at these judgments. More dangerously, you may conclude that you have nothing in common with some racially, religiously, or geographically defined group because of some habit or way of life they have. And, most dangerous of all, you may decide that, because of this habit or way of life, they are inferior. Nazi propaganda films about the Jews portrayed them as rats infecting society. They were filmed to look like that. And, since they had been stripped of their possessions and forced to live in ghettos, they looked pretty bad—anyone would. But it is difficult, sometimes impossible, for people not to conclude from the condition of those in poverty or profound suffering that this is somehow related to their ultimate nature.

We all repeatedly fall into the trap of identifying people too closely with their circumstances. It is an almost overwhelming human impulse to attach values to appearances. The habitual ease with which we do this explains a great deal about the history of eugenics and its extraordinarily seductive power as a theory.

With both these philosophical and psychological points in mind, I can now turn to the story of modern eugenics. We can start with Francis Galton, who was born the same year—1822—as Gregor Mendel and was also Charles Darwin's cousin. So, from birth, he seemed destined to play a leading role in the story of modern biology. Even before the momentous publication of his cousin's *Origin of Species* in 1859, Galton had been fascinated by the possibility of finding some material basis for assessing human beings. He had, for example, taken an interest in phrenology—the attempt to draw conclusions about human personality from the shape of the head. This idea is now utterly discredited, but one cannot help admiring its stolid Victorian literalness: the personality was in the brain, the brain was in the head, so the head's shape must reflect the type of brain within. Like *limpieza de sangre*, or like contemporary genetics, phrenology was one way of identifying a material basis for human variation. But phrenology was going nowhere; it had no serious experimental basis. So it was to be Darwinian evolution that really fired Galton's imagination.

For Galton, evolution turned the Christian interpretation of history on its head. The Christian story says we are fallen from grace and can only be redeemed by achieving union with God. But evolution seemed to show a gradual process of improvement here on Earth. Natural forces were at work improving organisms. Far from falling from a high state of grace to a low one, mankind was rising from the primordial soup to ever higher standards of perfection.

Galton's point of view can be seen in the starkly direct opening paragraph of his *Hereditary Genius*, published in 1869, ten years after Darwin's *Origin*:

> I propose to show in this book that a man's natural abilities are derived by inheritance, under exactly the same limitations as are the form and physical features of the whole organic world. Consequently, as it is easy, notwithstanding those limitations, to obtain by careful selection a permanent breed of dogs or horses gifted with peculiar powers of running, or of doing anything else, so it would be quite practicable to pro-

duce a highly-gifted race of men by judicious marriages during several consecutive generations. I shall show that social agencies of an ordinary character, whose influences are little suspected, are at this moment working towards degradation of human nature, and that others are working towards its improvement. I conclude that each generation has enormous power over the natural gifts of those that follow, and maintain that it is a duty we owe to humanity to investigate the range of that power, and to exercise it in a way that, without being unwise towards ourselves, shall be most advantageous to future inhabitants of the earth.

Like Darwin, Galton had moved from the artificial selection of animal breeding to wider conclusions about the history of life. Such a vision seemed both to offer and to demand an objective assessment of human qualities. The word "degradation" indicates that Galton believed, at one level, that he was capable of such an assessment and, at another level, that his project was based on the possibility of the much more detailed planning and refinement of this assessment. His central point was: If we work with rather than against the overwhelming natural force of evolution, then we should use our reason to help that force, to accelerate the process of improvement, and to avoid any signs of decay in the human species.

So it became a matter of scientific urgency to understand humanity. And one powerful tool was at hand—statistics. Galton was a statistician of genius, and his key insight was the way statistics could be used to arrive at generalizations about the human population. The science of statistics is the foundation of modern genetics.

Statistics are the secret weapon of the politician and the economist. They are, like the telescope or the microscope, a way of extending human perception. If, for example, I travel through a strange country and note the activities of various people as I come across them, I will end up with a random collection of observations which may be impressionistically interesting, but which will be of little practical use. There are people who work in offices, in the fields, and in factories. Some do no work at all. And that is that. But say I then employ armies of people—bureaucrats—to count exact numbers or to take representative samples of the population of this country. I then bring all these numbers together. Now I can make larger and more precise statements—say, that 30 percent of the workforce is in the fields, 40 percent in factories, 10 percent unemployed, and so on. The country exports so much and imports so much. Statistics have created new facts about this country, a whole new way of knowing.

Almost all our general knowledge of the world is now expressed in such statistics. The economic indicators that dominate the news, the analysis of election results, or even the nature of people's aspirations all appear in statistical form. Statistics are one of the great truth bearers of our time.

But this development is now so familiar that we sometimes do not notice its moral implication. For statistics create a realm of truth far above the level of the individual. Statistics produce images of great economic and social trends or movements and, in doing so, alter our perception of the individual. The worker in the field or the factory is no longer merely an individual; he is part of a larger, mathematized scheme of things. More than this, he is part of many different mathematical systems—as productive unit, as consumer, as one small element in a number of psychological and social aspirations, desires, or failings. In each of these he becomes part of a systematically simplified pattern. One specific aspect of his identity is placed on a graph or chart. If we are measuring, say, his height, then he will be placed on a graph which, for the population as a whole, will form a "bell curve," with the most common height forming the high point of the curve and those taller or shorter than the average distributed about this point in a form known as a "standard distribution."

This is all so familiar to us that it scarcely seems worth pointing out. Moreover, it seems harmless—merely a useful tool like any other. But note that a new way of considering the individual has appeared in the world, a way which cannot fail to have a moral impact. Suddenly the individual has become part of a mathematical collective. The rough edges of his personality have been isolated and then smoothed into the regularities of a bell curve, a bar chart, a pie diagram, or any of the other familiar pictures we use to explain statistics. Note, too, the direction of the moral change that this implies. It pushes us toward a view of the individual, not as Kant's absolute and final moral point, but rather as a small element in an overall pattern. We may still believe in the absolute moral fact of the individual, but it will be a subtly more difficult belief to sustain. We cannot help but see the individual as a part of something rather than as the untouchable end point of a process. It is but a short step from here to seeing the aberrant or abnormal individual as an inconvenience, a blot on the picture. It is often said that there are "lies, damned lies, and statistics." It should perhaps be added that there are tyrannies, murderous tyrannies, and statistical tyrannies.

Statistics became the foundation of eugenics and, later, of population genetics, which formed the basis of some of the most important breakthroughs in contemporary genetics. But for Galton, they were the objective key that unlocked the messy subjectivity of the human race. Armed with evolution and statistics, he conceived of a science that would chart the progress of the species, defining its place on the evolutionary ladder. He and his followers, notably Karl Pearson, insisted on the existence of an objective, biological truth about people to the point that Pearson argued that there was no point in improving schools since intelligence could not be taught; it could only be bred. At once this makes clear how radical the social consequences can be if we exchange a cultural for a biological view of humanity.

It should, at this point, be remembered that these ideas in the latter half of the nineteenth century were based on a combination of total ignorance and a willfully shortsighted interpretation of the mechanism of natural selection. Although Galton and Pearson could confidently point to the process of heredity, like Darwin, they had no idea how that process worked. Had they known what we know today about heredity, they may still have clung to the principle behind their ideas, but their project would have become substantially more complex. And their interpretation of Darwinism as a movement upward—a definite, detectable progress toward higher and better forms of the organism— was a wild oversimplification. Whether we can say evolution is progressive is a complex argument among scientists. But it is clear that it is a process full of dead ends, multiple species catastrophes, and, suddenly, fatally transformed environments. An organism that successfully adapts to one environment may find that, when the environment changes, it is hopelessly handicapped. And environments are constantly changing. If evolution is progressive, then this is a very inefficient and accident-prone form of progress.

Yet, in spite of these profound problems with the initial eugenic position, the central principle of eugenics seems so obvious that it must be true. It is this: The more the better and more clever people breed and the less the worse and more stupid people are allowed to breed, then the better and smarter the human species as a whole will grow. This is so because of the basic genetic principle that like produces like. This is also not exact. Good, smart people will sometimes produce bad, stupid children and vice versa. But, and this is the crucial point, statistics allow you to see the big picture. And the big picture is clear: usually good, smart people produce good, smart children, and usually bad, stupid people produce bad, stupid children. The exceptions are statistically insignificant. (I am not, I should say, making any judgments at this point about the meanings of "good" and "smart." I am merely using those words to make the eugenic issue as clear as possible. "Good" and "smart" simply stand for all desirable human characteristics, whether moral or intellectual. For different people they may mean different things.)

The reason this eugenic principle struck Galton and his followers with such force was that they were witnesses to a historic reversal in social structures. Prior to the industrial age, infant mortality was high, and it was highest among the poor. The rich could provide superior care, sanitation, and nutrition, and, as a result, a larger proportion of their children survived. And, since everybody, rich and poor, wanted large numbers of children, the absolute proportion of children born to the rich would, if anything, tend to rise. A Galtonian eugenicist would be happy that the human race was moving in the right direction.

With the rise of an urban, industrial society, a change began to occur. Populations as a whole increased. New social orders emerged in the cities. People became wealthier and a higher proportion of children survived. Meanwhile, for whatever reason, richer families started having fewer children. The eugenicist's dream became a nightmare. The human race faced "dysgeny"—the decline of the quality of the species. The fast-breeding urban masses represented a threat to the human future.

Yet again it is necessary to point out a crucial flaw in this early eugenic insight—though, again, it does not necessarily compromise the basic position. The psychological point I made earlier—that we tend to identify people too much with their circumstances—was all too clearly exemplified by the anxious insights of this first wave of eugenicists. They had a simple vision of society, based largely on the crude stratifications created by mass industrialization. There were, in their eyes, the masses and the elite. The elite tended to be a relatively small group who preferred to have fewer children. The masses were a vast group which reproduced profusely. To this day the social distinction between the reproductively prolific and the reproductively restrained remains. In the final episode of *Cheers*, the television sitcom, Diane, the hyper-refined heroine, reels back in horror when she learns how many times Carla, the working class barmaid, has given birth. "Good God," cries Diane, "you breed like a fly!" Breeding like a fly remains a clear indication of low social status—perhaps because it makes you look too much like an unthinking, unsophisticated victim of basic biology.

But, to the eugenicist, this lower-class fecundity was more than just an indication of social status; it was a threat to the future viability of mankind. It seemed that modern industrial society had created a system that would drive the human race back down the Darwinian ladder. But, precisely because the stratifications of nineteenth-century industrial society were so crude, the eugenic anxieties of the Victorians were overstated. The rigidity of social structures meant that the urban slums were bound to be full of unacknowledged talent and genius that simply could not find expression because they had no way of joining the elite. The masses the late Victorian elite contemplated with such dismay were not created by stupidity or moral delinquency; they were created by specific economic conditions at a specific historical moment. The elite had made the mistake of identifying people with their circumstances. In view of the crushing power of these circumstances, there was no sure way of concluding that the proportion of clever, upright people was any higher among the elites than among the masses. In other words, low social status did not provide any kind of evidence as to your badness or stupidity. It only, through the eyes of Galton and his contemporaries, looked that way.

This vision could only be reinforced by the rediscovery of Mendelian genetics in 1900. Here was a clear, beautifully simple mechanism for the transmission of hereditary characteristics. The eugenicists were delighted. The race was on to find as many clearly Mendelian characteristics in humans as possible. Here, at last, was the material source of inherited variation. Here, potentially, was the evidence needed to justify eugenics.

So persuasive was the argument that, whatever the shortcomings of its initial conception, the basic eugenic vision of the malign combination of a low-breeding elite and the high-breeding masses has ever since been an almost ineradicable element in social debate. Major figures—H. G. Wells, Bertrand Russell, George Bernard Shaw, Winston Churchill— have been convinced by its logic and persuaded of its urgency. Eugenics, as a result, became a powerful movement around the world, especially in Britain and America. To be a eugenics enthusiast was to regard oneself as committed to a no-nonsense, progressive, hard scientific view of humanity.

As George Bernard Shaw put it: "Being cowards, we defeat natural selection under cover of philanthropy; being sluggards, we neglect artificial selection under cover of delicacy and morality."

Note the form of this thought. Charity and idleness are threatening the quality of the human race. Our sentimental and caring side—our "philanthropy"—defeats natural selection, and, being fastidious about our social ideas and sexual standards, we lack the courage to force the human race to breed selectively. The message is: We should suppress our charity and our standards in favor of a more ruthless program of improvement.

And Bertrand Russell, perhaps the most politically influential philosopher of this century, spoke explicitly of the replacement of religion by science as a basis of morality. "The idea," he wrote, "of allowing science to interfere with our intimate personal impulses is undoubtedly repugnant. But the interference involved would be much less repugnant than that which has been tolerated for ages on the part of religion."

So eugenics allowed the elites in the early years of this century to sound both tough-minded and socially concerned. If they were not narrowly concerned about poor people, they were more broadly concerned about the human species in general. Eugenics also allowed them to be seen as rebelling against Victorian prudery. It was a subject that required open and frank discussion of sexual behavior, and this gave men like Shaw the opportunity to shock an older generation.

In Britain, eugenics was very much associated with left-wing politics. It was a "progressive" movement that seemed to fit well with the then still credible idea of a planned, socialist Utopia. The Soviet Union was still able to present itself to Western intellectuals as a great and successful experiment. And that society was based on the primacy of the masses over the individual, on the statistical rather than the specific truth. Such a conception allowed intellectuals to disregard the local inhumanity of a eugenic program in favor of the grander vision of a humanity that could rationally improve itself and create a happier, more humane future. In the United States, on the other hand, eugenics was associated with a more right-wing desire to reduce the cost of bad breeding. Eugenics propaganda made much of the tax burden imposed by the mentally retarded or the physically handicapped, people who, in a properly eugenic society, would never have been born. There was, for example, the famous case of the Jukes family, which included 18 brothel-keepers, 128 prostitutes, 76 convicted criminals, and 200 recipients of public welfare. The Jukes were discovered by New York City merchant Richard Dugdale in 1874 and were estimated to have cost the state more than $1.3 million in welfare and lost productivity.

The argument in the United States had a practical rather than idealistic basis, and, maybe because of that, it was the Americans who were most eager to embark on eugenic legislation. Or perhaps there was simply something about eugenics that appealed to the American imagination. Certainly the thoughts of Herbert Spencer were hugely influential in the United States. Spencer was an English Darwinian who saw natural selection as "the survival of the fittest," a moral injunction to support the best in any society and suppress the worst by biological means. He gave birth to the idea known as "Social Darwinism." This was taken, by American businessmen, to be a justification of their own activities. They could now see themselves as healthy expressions of beneficial, natural forces.

But, whatever the causes of the prewar American fascination with eugenics, the lesson is clear. Free, democratic societies are perfectly capable of talking themselves into savage behavior on the basis of a scientifically persuasive argument. So Charles B. Davenport, the founding father of the American eugenics movement, could say with a chilling combination of scientific jargon and barbaric emotion, "Society must protect itself; as it claims the right to deprive the murderer of his life, so also it may annihilate the hideous serpent of hopelessly vicious protoplasms."

Indeed, there were no lengths to which Davenport would not go in his effort to exploit science to subvert morality. For example, he contested the argument that America had always been a plural, immigrant society by appealing to Mendel. The idea of the nation as a "melting pot" had been rendered meaningless by Mendelian genetics, he claimed, because this new science showed that characteristics were inherited as units—they did not blend, they did not melt. One bad gene could infect the whole. Therefore, once admitted, bad characteristics would persist rather than being dissolved.

This was crude science, but it was believed. Evidence of the power of the germplasm (the term then used for the hereditary material)—and the possibility of its deterioration—was there for those who wished to look for it. During the First World War, for example, intelligence tests administered to American soldiers appeared to show that the mental age of the average white draftee was only thirteen years. And after the war, the feeling grew in both scientific and lay circles that the new immigrants were not up to the quality of the old. Unless something was done, America would be stuck with dysgenic decay.

Something already had been done. Inspired by eugenic anxieties, some states had, before the war, imposed compulsory sterilization laws. By the late 1920s a total of two dozen sterilization laws had been passed. Then, in the 1920s, immigration was seriously restricted—a momentous decision that was to determine the future racial structure of American society.

The detailed history of American eugenics has been written elsewhere, notably by Daniel J. Kevles. It is not a story of which Americans can be proud. Scientific conviction was exploited to justify massively inhumane and divisive policy. It may seem incredible to us today—as remote and wrongheaded, perhaps, as the Spanish Inquisition's pursuit of *limpieza de sangre*. But the truth is that this was little more than sixty years ago.

It was a time when Western society was first beginning to face the profound, world-changing force of the new biology created by the unification of Darwin and Mendel. This force was ambiguous. It seemed to offer limitless benefits, yet it would only deliver, as the English Catholic writer G. K. Chesterton pointed out, if we overturned our traditional moralities. So looking after and preserving the rights of a handicapped baby may once have been seen as a saintly act; in the eugenic world, it became destructive—an act that encouraged the genetic deterioration of the species. If we wanted to improve the human race, it seemed we had to become worse people.

One English writer captured this paradox exactly. Aldous Huxley felt both the seductions and dangers of technology. He saw that if the technology worked, who could argue? What reason could there possibly be for not going with the technological flow? In 1932 Huxley published his novel *Brave New World*, set in a future society in which total eugenics has finally triumphed. Breeding is carried out artificially and according to a careful plan that precisely determines each individual's place in the world. Usually the novel is seen as a simple condemnation of this eugenic society, but it is not. Huxley saw its possibilities, its freedom from conflict, as clearly as he saw its shortcomings.

In the book, the Savage—a human bred by ordinary means—appears as an exotic curiosity in this brave new world. He confronts the Controller, Mustapha Mond:

> "But I don't want comfort. I want God, I want poetry, I want real danger, I want freedom, I want goodness, I want sin."
>
> "In fact," said Mustapha Mond, "you're claiming the right to be unhappy."

The book, and indeed the whole eugenic-genetic debate, pivots on that wonderfully distilled moment of confrontation. The Savage wants the full, passionate range of human experiences. But why bother, asks Mond, if you can simply be happy? Those experiences are, after all, little more than grandiose expressions of the problems of life, and these problems can be eliminated.

This was the crucial confrontation, defined in the 1930s when ambiguous visions of the technological future—communist, fascist, or malevolently capitalist—first began to surface. Such visions had provoked fear of the eugenic ideal. The eugenics movement itself had, by this time, begun to soften its arguments. Reform eugenics, as Daniel Kevles calls it, tended to take the less harsh line that improving social conditions would inspire the masses to have fewer children. There should be no need either for sterilizations or for any other coercive policies. But the central logic of reform eugenics remained that the human race was faced with genetic deterioration unless we actually intervened in reproductive decisions. The cultivation of nature, not the improvement of nurture, was the way to the future. All the nurture in the world could do nothing to save the imbecile or redeem the congenitally afflicted.

A Geneticist's Manifesto—signed by 22 British and American scientists in 1939—called for the replacement of the "superstitious attitude towards sex and reproduction now prevalent" with "a scientific and social attitude" that would make it "an honour and a privilege, if not a duty, for mother, married or unmarried, or for a couple, to have the best children possible, both in respect of their upbringing and their genetic endowment." The great British biologist J. B. S. Haldane produced a novel—*Daedalus*—in which he insisted that the technological inventor was a Prometheus whose every innovation was destined to be "hailed as an insult to some God." Biology for Haldane was the greatest of all contemporary innovations; inevitably it would face resistance from the scientifically ignorant, but, in time, it would come to be seen as the best—the only—way forward. This did not sound like the anxiety-laden, negative eugenics of Davenport; it sounded like something fresher, newer, less sinister, more exciting.

But, either way, it hardly mattered what Haldane or any of those 22 scientists were saying in the thirties. It hardly mattered what all those eugenically inclined state governments thought. For, in Germany, another writer had produced a book that took up the noble cause of eugenics, a book that was to be more effective than anything Galton, Pearson, or Davenport ever produced.

"Whoever is not bodily and spiritually healthy and worthy," wrote the author, "shall not have the right to pass on his suffering in the body of his children. . . ." The book was *Mein Kampf* and the writer was Adolf Hitler. The science of genetics was about to conduct its most spectacular experiment.

chapter 4

Eugenics 2: Tattooing Foreheads

I'd like to believe you can't do science if you're a really
evil person, but it's probably a romantic view.
—PETER GOODFELLOW, geneticist

Consume my heart away; sick with desire
And fastened to a dying animal
It knows not what it is. . . .
—W. B. YEATS

Ernst Haeckel was a central figure in nineteenth-century German
biology. He had a theory of nature which he called "monism." This
was turned into a movement led by an organization known as the
Monist League. Monism was simply the opposite of dualism. In dualism there is a material world and a spiritual world—the world of mind.
In monism, the world is a single, material unity, with no special spark of
life. Monism was yet another version of the antivitalist program, or the
scientific war on the idea that there was something special about living
matter.

But Haeckel and the monists went further than this. They insisted
that the basic unity of nature meant that humans were just like other
animals. Human self-consciousness and its by-product, human culture,
were merely quantitative improvements on the achievements of animals. There was no fundamental, qualitative difference, no ultimate
discontinuity.

Thus far Haeckel may not appear especially revolutionary. Many de-
cent, reasonable people would agree with both these points, though a
few might dissent from the materialist view and many more would dissent from the idea that humans and animals are much the same. But

these are well within the range of normal, arguable positions. In the wrong hands, however, normal arguable positions can surprisingly easily turn into barbarism.

Adolf Hitler liked the theories of Haeckel and the Monist League. He seems to have had a good grasp of the science involved. In fact, Hitler's monist-influenced book, *Mein Kampf*, has, as the biologist John Vandermeer has pointed out, a surprising degree of biological credibility. Hitler wrote of the way subpopulations of species form through geographical isolation—a perfectly respectable viewpoint. He wrote of the way animals in these subpopulations gradually formed separate races—a reasonably respectable viewpoint, though biologists would not now use the word "races." If these subpopulations then came together, one would either destroy the other or they would intermingle to produce an intermediate "race"—a touch simple, but a viewpoint still within the realm of respectability.

But the mask of respectability fell away when Hitler applied the same thought processes directly to human populations. Human culture, he pointed out, made a difference to these basic processes. When one superior human subpopulation encountered an inferior subpopulation, the latter was not destroyed, it was enslaved. The two populations, therefore, coexisted, and this led inevitably to interbreeding and a "mongrelization" of the superior race. In employing slavery rather than slaughter, human culture was thus working against the evolutionary improvement of the species by mingling the hereditary material of the superior races with that of the inferior. This was straight from Haeckel.

"The artificial selection practiced in our civilized states," Haeckel wrote, "sufficiently explains the sad fact that, in reality, weakness of the body and character are on the perpetual increase among civilized nations, and that, together with strong, healthy bodies, free and independent spirits are becoming more and more scarce."

This is an insight that was not so very different from the dysgenesis glimpsed by Galton and his followers. But they, perhaps, would not have taken the next step. Haeckel—and, subsequently, Hitler—reasoned that the gap between the most civilized races and the lowest savages was greater than the gap between the savages and animals. Therefore, the lives of savages were of much lower value. They could be disposed of in whatever way the superior races chose. Above all, the superior races must purify themselves in the name of human evolution and the inferior must be suppressed, prevented from tainting the slowly purifying blood of their betters.

Most ideas—even powerful and effective scientific ideas—spend much of their time in a kind of hibernation. Sometimes this is because they are ideas that, however penetrating, do not seem to have any direct effect on the world. Sometimes they are waiting for the technology which will make them effective. Sometimes their effects are pervasive but concealed by the clamor and events of history. Sometimes they are simply wrong and do not work, however convincing they may seem to their adherents. But sometimes ideas leap straight out into the world. One such idea was Marxism; another was Nazism. Perhaps it was the sheer simplicity of these two ideas that made them so instantly effective. The first simply stated that history is destiny, the second, that biology is destiny.

Nazism used biology to justify mass murder. In doing so it detonated biological complacency, shattered public and scientific confidence, and tainted the entire subject, especially genetics, for years afterwards, possibly forever. Some would deny this. Recently, the British geneticist Steve Jones wrote: "Genetics is, at last, like Germany, ready to stop apologizing for its past." But this shows a failure to understand the true depth of the problem revealed by Nazism. For Hitler demonstrated the appalling dangers implicit in science's invasion of the human realm.

At this point scientists will become angry and argue that Hitler's science was not really science at all. Even when it wasn't completely wrong—which it was most of the time—it was absurdly oversimplified, they say. To condemn a science on the basis of one psychopath is like banning baseball because a bat was once used to murder somebody.

But this is to miss the point completely. It is a self-serving argument that can be used to free science from any responsibility at all. I believe it is essential that this argument be rejected. And here, in the contemplation of the catastrophe of Nazism, is a good place to do it.

First, as I have said, much of the basis of Hitler's biology was valid in the light of knowledge at the time. It was arguable but scientifically respectable. We may say it is wrong now, but we could say the same of most past science. For example, we can say Newton was "wrong" because Einstein's theories showed that his celestial mechanics were, in reality, a rough generalization, not "true" in any absolute sense. In these terms most science is wrong most of the time—not just the "bad" science of Haeckel, but also the "good" science of Newton. Five hundred years from now, virtually all the science of our day will be regarded as similarly wrong. It is, therefore, ridiculous to say that the institution of science is unaffected by the mistakes of the past because they were based on wrong science.

Second, it was the very persuasiveness, the immense authority of science that convinced Hitler and his followers. For, in spite of being wrong most of the time, science always *appears* to be right. Indeed, it increasingly appears to be the only right thing available. This is in part because scientists always say they are right, but it is also because science is so extraordinarily effective. I cannot, by reading the Buddha's Fire Sermon or reciting Christian prayers, make a Boeing 747 fly from London to New York. By applying engineering and aeronautics, I can.

This staggering effectiveness convinces people that science is all-powerful and that if something is labeled "scientific," it must be true or feasible. Scientists, of course, go along with this because it exalts their social status. Yet, in doing so, they are implicated in the outcome—whether it is pollution of the environment by some insecticide or the murder of six million Jews. Science is guilty in both cases because scientists said these things could or should be done. It provided reasons that were believed because science was believed to be the truth. The message should be clear: To perpetuate any gospel of the omnipotence or even the omniscience of science is to dip your hands in blood.

Third, the application of science to the human realm is always going to be fraught with dangers, no matter how well-meaning the scientist. Establishing truths of human behavior is difficult because, on the one hand, the scientist is human; he has attitudes and opinions. As the history of eugenics shows, scientists invariably have a cultural bias which they apply to their observations. On the other hand, the human observed is changed in unquantifiable ways by the act of observation. When told, for example, of the existence of the subconscious by a Freudian psychoanalyst, you become a different person; in ways that are impossible to measure, your behavior will be changed. This makes it all but impossible to be sure of any generalization about human behavior.

We should be warned by the crass absurdity of the supposedly scientific observations of humanity in previous generations. For example, look at phrenology. We cannot assume we are so much smarter than the phrenologists. James Watson and many other scientists have argued that we will not make the mistakes of the past because we know so much more. That is absurd, first because we will never be able to judge when "more" becomes *enough* and, second, because there is no reason to suppose that even perfect scientific information will make us behave any better.

And, finally, science tends toward radicalism. Science is reductionist. Its success is based on the assumption that the whole can be explained in terms of its parts. It aspires to discover the ultimate simplicity behind complex phenomena. Some scientists today are challenging this approach, but it is unarguable that the history of science is the history of reductionism. The conclusions of any reductionist process will inevitably appear radical. It is, for example, a radical idea that the table on which I am writing this consists largely of empty space. But it is, scientifically speaking, true. Marxism and Nazism were both based on the reductive radicalism of science. Just as the physicist says, against all the common sense evidence to the contrary, that this table is largely empty space, so the Marxist says all human truth is history, and the Nazi says all human truth is biology. In the case of the table, the effect of this radicalism is only to give me a passing feeling of vertigo; in the case of Marxism and Nazism, the effect is to kill millions. In addition—and this is a crucial point—even if some ideology were to come along based on immaculate science rather than the dubious speculations of Marx and Haeckel, it may still kill millions. Science is under no obligation to produce morally acceptable outcomes. It would be best and safest, therefore, for us all to agree on its incompetence in the human realm. To quote that great historian of ideas, Sir Isaiah Berlin, "To claim the possibility of some infallible scientific key . . . is one of the most grotesque claims ever made by human beings."

Tell the world that there is some scientific basis for differences between people—that there are "good" and "bad" genes—and, as night follows day, somebody somewhere will use it as an excuse to start killing people. No geneticist working today can ignore this fact. All of which is to say that Steve Jones is quite wrong to imply that genetics can break free from its Nazi past. Neither Nazism nor Marxism were passing aberrations; they were catastrophes whose precise natures were defined by science. Clearly there have been other catastrophes that have had nothing to do with science. And I am not saying that science caused either of these appalling movements; obviously there were many other factors involved. But science was deeply implicated and, in both cases, provided the rationale for slaughter. Science is implicated in the wars of the twentieth century much as religion was in the wars of the Middle Ages. Indeed, science is implicated more if we include the fact that science provided the advanced weaponry that made our modern wars so destructive. To deny that or to forget it, as Jones suggests, is to risk making the twenty-first century as bloody as the twentieth—a century which has the dubious honor of being, so far, the bloodiest in human history.

In the specific context of my narrative, Nazism was a phenomenon that seemed deliberately designed to fling the arguments of the eugenicists back in their faces. Auschwitz and Dachau revealed the truth behind the German eugenic practices that so many Americans had admired so much. As G. K. Chesterton argued so passionately in his book *Eugenics and Other Evils*, the very idea of eugenics was based on the making of a moral distinction between people that went directly against Christianity and, though Chesterton did not make this claim, against the crucial Kantian ideal of the Enlightenment—the individual as moral absolute. Eugenics places the good of the species and its future far above the rights of any individual. Nazism took this further by elevating not the whole species but one particular race—they called themselves the Aryans—to the pinnacle of creation. What was good for this race dwarfed all other considerations. And, the Nazis decided, what was particularly good was killing Jews because eugenics demanded they be permanently prevented from tainting the *limpieza de sangre* of the Aryans.

After the war, witnesses from Auschwitz revealed that Nazi doctors had been testing a variety of different methods of sterilization—castration, X-rays, injections, electrocution of the genitals—that would enable them to sterilize inferior races and repopulate all Western European countries with Germans within one generation. The eugenicists suddenly had a lot of explaining to do.

At first, they did not get much chance. The eugenics movement came close to annihilation as a result of Nazism. In 1947 one American eugenic society recorded, with superb understatement, that "the time was not right for aggressive eugenic propaganda." There was no way, after that experience, that people could continue to talk blandly of improving the species.

The whole intellectual climate changed. The hope for the future was suddenly seen to lie in the improvement of nurture rather than the enhancement of nature. Social progressives no longer spoke, as Shaw and Russell had done, of improving the population by controlling reproduction; rather, they spoke of investing in education and eliminating poverty. The fifties were a period of rapid economic growth and huge technological optimism, but the technology was seen as beneficial only when out in the world, not when meddling with the biological condition of the species. This cultural battle between nature and nurture is a subject to which I shall return.

The political scientist Diane B. Paul has illustrated one aspect of this change through three legal cases. The first was the pre–World War II Carrie Buck case, in which a woman was sterilized to prevent her from producing more "imbeciles." State-enforced eugenics overruled any appeal to individual rights. Then there was the postwar Karen Ann Quinlan case, in which the decision to terminate life-support systems was delegated entirely to family and physicians. Here only a private decision is seen as legitimate. And, finally, there was the 1981 Grady case, in which parents of a Down's syndrome girl wanted her to be sterilized to prevent her becoming pregnant as a result of rape or seduction in a sheltered workshop. The New Jersey Supreme Court ruled in the parents' favor, but insisted that all such sterilizations would require judicial approval. This was not, as in the Quinlan case, a private matter. The court's reasoning was inspired by the "sordid history" of eugenic sterilization.

But in fact eugenics had only temporarily gone into hiding. The eugenic idea survived the war remarkably intact. Ruth Hubbard and Elijah Wald point out that as late as 1941, when the nature of Nazism was becoming quite clear, the British biologist Julian Huxley, brother of Aldous, was firmly insisting on its solid, scholarly basis. "Eugenics is running the usual course of many new ideas," he wrote. "It has ceased to be regarded as a fad, is now receiving serious study, and in the near future will be regarded as an urgent practical problem." Huxley added that society must ensure that mental defectives do not have children.

This prewar idea persisted. After all, it had not been proved wrong; it had merely been wrongly applied. Eugenics was embarrassed but not refuted by the concentration camps. Meanwhile, the dysgenic anxiety and the apparently obvious solution—eugenics—remained powerfully persuasive to its many enthusiasts. But a new dimension was needed to restore eugenics to the center of respectable debate, a dimension which could free it of all Nazi overtones.

In part, this new dimension was simply scientific triumphalism. In the late sixties, as I have said, there was a climate of anxiety about the path technology in general was taking. But this anxiety was far from universal. Some were overcome with the majesty of the scientific project. In 1969, the National Academy of Science (NAS) produced a survey report entitled "Biology and the Future of Man." This was its final paragraph:

> Man's view of himself has undergone many changes. From a unique position in the universe, the Copernican revolution reduced him to an inhabitant of one of many planets. From a unique position among organisms, the Darwinian revolution assigned him a place among the millions of other species which evolved from one another. Yet, *Homo sapiens* has overcome the limitations of his origin. He controls the vast energies of the atomic nucleus, moves across his planet at speeds barely below escape velocity, and can escape when he so wills. He communicates with his fellows at the speed of light, extends the powers of his brain with those of the digital computer, and influences the numbers and genetic constitution of virtually all other living species. Now he can guide his own evolution. In him, Nature has reached beyond the hard regularities of physical phenomena. *Homo sapiens*, the creature of Nature, has transcended her. From a product of circumstances, he has risen to responsibility. At last, he is Man. May he behave so!

Note the way "he can guide his own evolution" joins the list of other scientific and technological achievements; this was written just before the emergence of recombinant DNA technology. And note the penultimate sentence; only by exerting total control over nature does the human being become Man.

Also from the late sixties, here is the geneticist Robert Sinsheimer explicitly evoking the possibility of a new, respectable, and socially just eugenics:

> The old eugenics was limited to a numerical enhancement of the best of our existing gene pool. The new eugenics would permit in principle the conversion of all the unfit to the highest genetic level.... It is a new horizon in the history of man. Some may smile and feel that this is but a new version of the old dream, of the perfection of man. It is that, but it is something more. The old dreams of the cultural perfection of man were always sharply constrained by his inherent, inherited imperfections and limitations.... To foster his better traits and to curb his worse by cultural means alone has always been, while clearly not impossible, in many instances most difficult.... We now glimpse another route—the chance to ease the internal strains and heal the internal flaws directly, to carry on and consciously perfect far beyond our present vision this remarkable product of two billion years of evolution.

I cite these two quotations here to indicate that postwar eugenic thinking has not always been hesitant and self-conscious in the shadow of the Nazi past. They also exemplify an important scientific view which, these days, is not often so frankly stated. Both Sinsheimer and the NAS report took the view that biological control represented a logical extension of mankind's glorious progress to complete mastery of nature. This progress is seen as the defining project of the species. In both quotations one feels that nothing else but science is of significance. All other human activity and achievements is reduced to a sideshow, for it is science that changes "man's view of himself," and it is science that can "ease the internal strains and heal the internal flaws." If science alone can do all these things, it is difficult to know what else there is to be done and what possible historical status can be accorded to anybody who is not a scientist. And, most important for my discussion of eugenics, it becomes inconceivable that there could be any argument for not eugenically controlling human reproduction; it is merely one more aspect of our entirely virtuous progress to the total control of nature.

But there was also a specific scientific development that was to make eugenics respectable again: molecular biology. From 1953 onward biology acquired startling new forms of knowledge. Just as sentiment in the fifties had been moving away from nature toward nurture, precise biological information—information about nature—was suddenly becoming available. And this information appeared to be clear, deterministic, and possessed of a precise mechanical structure. It defined that previously theoretical entity, the gene, and it revealed the elementary, numerical nature of the code through which it worked. Admittedly this apparent simplicity was, in some sense, illusory in that the biochemical pathways by which genes express themselves have turned out to be fiendishly complex. Nevertheless, the fact that there was some ultimate form of simple order to life seemed to endorse some of the apparently commonsense insights of the eugenicists. So, while nurture took over a public world revolted by the naturalism of the Nazis, nature was quietly taking over the semiprivate world of science.

The key point about molecular biology as far as the eugenicist is concerned is its hard materiality. It locates diseases, traits, and personalities at precise points within a known chemical structure. It does not get bogged down with the vagaries and seemingly infinite complexities of sociology or psychology; it provides, instead, precise numerical and physical information. So a whole range of human characteristics may, ultimately, be revealed by molecular biology to be traceable within the material realm rather than unknowably dispersed in the indefinable ether of the soft sciences. And that may soon give us the power to take direct control and improve the species.

Yet molecular biology grew up in the shadow of Nazism. The war had ended only eight years before Watson and Crick deciphered the structure of DNA. However much this new science might seem to support the insights of the eugenicists, the moral stigma was still too fierce. What was needed was a new moral dimension, a new justification for science to act in the human realm.

Disease supplied the moral dimension. Disease was the crucial factor that readmitted eugenic thinking to the respectable world. The suffer-ing caused by disease could morally justify interventions that either verged on or were explicitly eugenic. The starting point was the in-creased understanding of the devastating single-gene disorders which, from the seventies onward, were being traced to specific locations within the genome. The importance of these conditions was that they tended to be dramatic, easily trackable by basic Mendelian methods, and prone to cluster within certain human populations. Cystic fibrosis, for example, occurs mainly in European whites, while sickle-cell anemia occurs among blacks. But once we knew all this, what could we do about it? Merely knowing where the gene was offered no immediate prospect of cure. One very distinguished scientist, Linus Pauling, who had first identified the form of hemoglobin associated with sickle-cell anemia in 1949, had the answer:

> I have suggested that there should be tattooed on the fore-head of every young person a symbol showing possession of the sickle-cell gene or whatever other similar gene, such as the gene for phenylketonuria, that has been found to possess in a single dose. If this were done, two young people carrying the same seriously defective gene in single dose would recognize this situation at first sight, and would refrain from falling in love with one another. It is my opinion that legislation along this line, compulsory testing for defective genes before mar-riage, and some form of semipublic display of this possession, should be adopted.

This is an extraordinary quotation. It captures in just a few words how brutal eugenic impulses can so easily be resurrected. Disease pro-vides the excuse, and we are at once back to the fiercely coercive ideas of the twenties. If it proves nothing else, Pauling's quote illustrates how dangerous it would be to give scientists control of social policy.

But it does prove much more. Pauling's logic is untouchable. Sickle-cell anemia is a recessive single-gene disorder. Everybody has two copies—two alleles—of most of their genes. Because it is recessive, one copy of the sickle-cell gene will be canceled out by the dominant, nor-mal version of this particular gene. With one copy you will not contract the disease; with two copies, you will. Therefore, if two people, each with a single copy, marry, they will risk, on average, one in four of their children having the disease. What could be more sensible, therefore, than to ensure that such people do not take the impractical step of falling in love?

Yet the logic leads directly to a clear eugenic position—the control of reproduction, in this case by the preposterously public means of a sign on the forehead. In practice, this can be—and has been—more discreetly done. Tay-Sachs disease, for example, is a single-gene disorder that occurs among Ashkenazi Jews. It destroys the insulating sheath around nerves and leads to death in infancy. In the Lubavitcher community of Orthodox Jews in Brooklyn, a novel method was found for fighting this disease. All children were tested anonymously and the results recorded secretly. If two people who wished to marry were both carriers, they would be told not to marry (marriages are often arranged in this community, so this need not necessarily be an odd injunction). If one was a carrier and one wasn't, there would be no danger of transmitting the disease to the children, so the marriage was permitted with neither partner knowing whether either of them was a carrier not. As far as possible, any stigma attached to the disease was avoided.

But however humane the method, the intention is eugenic; indeed, the organization that promoted this way of combating Tay-Sachs is called the Association for an Upright Generation, a eugenically inspired title. This particular procedure may not be objectionable, but it shows the way in which certain forms of knowledge change, however imperceptibly, moral and social barriers.

And this change can be brutal. The Brooklyn approach to Tay-Sachs was a model of discreet care compared to the fantastic insensitivity displayed by the administrators of the sickle-cell screening program in the United States in the mid-1970s. As I have said, sickle-cell anemia is a recessive single-gene disorder, common among blacks; about one in twelve blacks in the United States are carriers. A number of states required by law that schoolchildren be screened for the gene before they could enter school. On the face of it such screening seems a reasonable aspect of a benign attempt to combat disease. And, indeed, the Black Panthers, a radical political group active at that time, regarded it at first as a sound health measure for black communities. The problem, however, was one of ignorance and an appalling lack of planning. What was not made absolutely clear was that possessing a single copy of the gene did not mean you were, in any sense, ill. Indeed, you might be said to be fortunate in that a single sickle-cell allele provides protection against malaria, which is why the gene had been concentrated in black and Mediterranean populations in the first place. As a result of the widespread ignorance of this point, blacks found to be carriers were discriminated against in jobs and insurance. They were particularly excluded from work in any job associated with flying because, it was thought, they could not cope with lowered air pressures.

It might simply be said here that the voluntary Jewish screening was well done and that the compulsory black screening was badly done. The lessons have been learned. But there is a deeper point which brings me back to the respectability or otherwise of contemporary eugenic thinking. The genetic identification of diseases provides physical evidence of a fundamental difference between races or ethnic groups, an idea that was previously taboo. Both Tay-Sachs and sickle-cell anemia are concentrated in minority populations. Cystic fibrosis, another devastating single-gene disease, is concentrated in a racial group—whites—but, for whatever reason, it has not received the same kind of publicity. As a result, it is seldom seen as a "racial" disease.

At the lowest level of debate, the first effect of the revelation that some diseases are racially or ethnically localized is to confirm to the already prejudiced their views about the inferiority of blacks or Jews. They are seen to be uniquely afflicted with exotic and appalling diseases.

But, at a much higher level, disease provides clear, physical evidence of differences between peoples. Even if we remove any specific stigma from the idea—by, for example, pointing out that cystic fibrosis is also a racially localized disease—the new genetic understanding of these diseases still seems to say that people are, indeed, divided by more than just appearances. This amounts to a refutation of the antieugenic nurture argument that says we are all the same beneath the skin and only environment creates differences. Clearly, if each of these dreadful diseases strikes specifically at different populations, then there is something more profoundly different than mere appearance. Beneath the skin we are not all the same.

A geneticist might respond by saying the differences are tiny—a matter of a few DNA base pairs among three billion. Yet the effects are enormous, and it is those effects that provide convincing evidence in the real world. As the Berkeley sociologist Troy Duster remarked to me: "People say it's at the molecular level, so then they will ask: What else is at the molecular level?"

Meanwhile, the sixties and seventies had seen the appearance of a new fear that was also, in essence, biological—the fear of a population explosion. Rapid global population increases combined with the new environmental anxieties of the time to create an image of an overloaded planet. There were simply too many people doing too much. Cutting down how much they did, and their wealth, seemed desirable but implausible. Reducing their levels of reproduction, given the new birth control technologies that had become available, appeared to be a much more practical option.

In 1971, Bentley Glass, the retiring president of the American Association for the Advancement of Science, took this logic one step further by arguing that the newly perceived need to ration children demanded a clear eugenic approach: "In a world where each pair must be limited, on the average, to two offspring and no more," he said, "the right that must become paramount is . . . the right of every child to be born with a sound physical and mental constitution, based on a sound genotype. No parents will in that future time have a right to burden society with a malformed or mentally incompetent child."

The language employed by Glass demonstrates a crucial development in the rhetoric of eugenics. Fear of a population explosion had created one of the curious moral transformations that Chesterton, prescient as ever, had foreseen. With humanity threatened by its own fecundity, morality now became solely dependent on an unborn future. From now on we must always, in our reproductive acts, be aware of the consequences for the resources of the planet. As Chesterton put it: "Now the eugenic moral basis is this; that the baby for whom we are primarily and directly responsible is the babe unborn." The future becomes a moral imperative, diminishing the status of the present.

In this case an even more radical moral transformation was involved. Once the limitation on the right to have children came into being, Glass suggested a new kind of right should spring into being—the right to a "sound genotype." Since we cannot have all the children we want, we have an obligation to ensure that those we do have are biologically fit.

Of course, strictly speaking, Glass's idea is absurd and incoherent. If such a right exists for limited numbers of children, why not for unlimited numbers of children? It is a peculiar kind of right that emerges only in times of reproductive rationing. More important, you cannot speak of the right of a child to a sound genotype. If it had an unsound genotype, it would, in his own terms, be a different child. Therefore, to claim the right to a sound genotype is like saying I have the right to be Michael Jordan. The concept is a ridiculous abuse of the idea of a right.

Nevertheless, the illogic in the position is revealing. What Glass is actually saying is that parents have a duty to bear only those children that are genetically sound. In order to support this idea, he invents this crazy category of the rights of the unborn. Once born, these future people will be able to assess whether or not their parents abused their rights by examining the qualities of their own genotype. The logic of such a position is that a fetus with an unsound genotype has a right not to be born. Muddying these waters with the language of rights results in the madness of giving a person who does not yet exist the right not to exist.

This is silliness, but it points to a profound problem at the heart of all eugenics and, indeed, genetics. If we are to stick to the Kantian or Christian sense of the absolute moral status of the individual, then we need a stable sense of what the individual is—where it starts, where it ends, and how it is constituted. For most of human history the concept of the individual has, indeed, been stable. The problem with contemporary genetic knowledge is that it threatens this stability. It seems to show us that each person is no more than a momentarily unique combination of its mother's and father's DNA. It makes us aware of an overwhelming quality of contingency at the heart of our existence; we are chance ripples in the gene pool. As Harvard biologist E. O. Wilson puts it, "The individual is an evanescent combination of genes drawn from this pool, one whose hereditary material will soon be dissolved back into it."

This creates the peculiar moral situation that Glass clumsily attempted to confront. For, in the light of genetics, it can seem strange to speak of an individual at all. Rather we become aware of an incomprehensibly vast, though finite, numerical field of possible DNA sequences. Any individual is simply one mathematical possibility among trillions of others. He becomes a statistical incident, surrounded by the ghosts of other possible incidents that did not occur, but which equally well might have. We become aware of the individual not as an absolute but as one version. There could, therefore, be better versions.

So it is the gene pool that has become the collective good for postwar eugenics, just as fewer tax dollars spent on imbeciles, or the future intellectual health of the species were the collective goods for prewar eugenics. In the simple but large-scale mathematics of molecular genetics, eugenics is reborn.

At this point I should acknowledge that many would argue that what I have been calling eugenics in the postwar period is not eugenics at all. Eugenics, according to them, involves coercion and a social aim; it must be a planned attempt to alter the human gene pool. So compulsory sterilization of people defined as criminal or mentally unfit is clearly eugenic in that it is coercive and socially targeted. But if we remove coercion—say, we merely ask these people to be sterilized—is it still eugenics? And what happens if we change the justification? What if we say that instead of preventing crime, it is to prevent a family from having a child with a serious disease? In the latter case we could say that not having a severely diseased child might be seen as a social as well as a purely private good. But what if—though this cannot yet happen and might never happen—a couple is told their unborn baby is going to be homosexual and they decide to abort? Is that eugenics? Or if they are told it will not have blue eyes and they decide to abort? Or is abortion itself—now massively practiced throughout the world—a form of eugenics? One leading geneticist told me it wasn't, because it didn't alter the gene pool. But an evolutionist would have to say it does, for abortion reduces the number of births to people who are prone to have abortions, and being prone to have abortions must, by the logic of the hard evolutionist, be a trait with a genetic element. The gene pool is thus altered.

For me it is all too obvious that those who wish to deny the title eugenics to anything other than coercive, socially targeted control of reproduction are doing so because they wish to avoid the Nazi taint.

They are restricting the use of the word so that any other form of reproductive intervention looks clean. And, to a limited extent, they have a point. Whatever your feelings about aborting all Tay-Sachs births or preventing them prior to conception, it is clearly an entirely different impulse from the Nazi's desire to liquidate European Jewry. But the attempt to eliminate the word is also an attempt to stop us from thinking. Eugenics is simply about the production of better human off-spring. This means that a huge range of human decisions—from deciding to have fewer children so that you can lavish more care on each of them, to compulsory sterilization—can be classed as eugenic. Clearly, to most of us, limiting the numbers of our offspring seems like a good idea while compulsory sterilization does not. We accept, without using the word, that there are good eugenic practices and bad ones. The debate should not, therefore, be blurred by concealed fears of the word itself; it is about where good eugenics shades into bad, and we can only make that judgment on the basis of our total view of human life.

The science of genetics makes this issue almost unbearably urgent because it places in our hands a far wider and more precise range of eugenic tools than ever before. We already have the power to detect prenatally a large number of disorders, though all we can usually do with that information is offer an abortion. Soon we will be able to detect many more disorders, as well as a number of "predispositions"—to heart disease, cancer, or whatever—yet it is improbable that we will be able to offer anything but abortion for many years.

Ultimately, of course, it is likely that we may be able to intervene genetically to alter the baby's genome so as to eliminate the disorder or predisposition. This specter of "designer babies" is often scoffed at by scientists as a science fiction idea that lies impossibly far in the future. I cannot see why they scoff. Many people abort fetuses found to be suffering from one of the currently detectable abnormalities and then later give birth to a child that is free of the disorder. This may be a primitive form of design, but it is design nonetheless, and all scientists know that the future holds an effectively infinite series of refinements of such choices. Like the word "eugenics," the phrase "designer babies" seems to alarm geneticists. But again, it is not the name given to something that is most important (in both cases these names are perfectly legitimate for a far wider range of practices than is normally allowed); rather, it is the scale of value we apply—where we decide good becomes bad—that matters.

So how would we make such a decision? In order to understand that question it is necessary to emphasize the conflict at the heart of all eugenic practices—the conflict between the public and the private, which is just another way of describing the conflict between freedom and social order that lay behind the Scopes trial and underpins the pornography debate today.

I asked James Watson about eugenics, and he began by admitting at once that private decisions are eugenic. "We do it right now in one sense," he said. "I wouldn't marry a stupid woman."

But then he went on to emphasize the immense significance of the shift from nurture to nature, inspired by molecular genetics:

> It would be nice if we were all equal. The Nazi viewpoint was so grim and awful that societies have since acted as if society didn't suffer from genetic inequalities. They've acted on that basis for the last fifty years—as if we can take care of everything through nurture. That's not going to be the case.
>
> I would like the answer to be that everyone is equally intelligent, but I think the answer we are going to find over the next twenty-five years is that people are genetically predisposed to mental diseases or to be rather stupid. It's not a nice answer and all we can do in society is work in the best way we can to reduce inequality.

Watson is not here advocating any form of state coercion; rather, he is simply saying that it is the provision of information that matters. He does not wish to remove decision-making from individuals.

One specific and very sensitive issue is homosexuality. We currently do not know whether or to what extent homosexuality is genetic. Dean Hamer, an American geneticist, published a world-famous study in 1993 that showed some linkage between homosexuality and a small region of the X chromosome known as Xq28. This was not, as the publicity tended to suggest, the "gay gene." It was not even a gene, but part of a chromosome that may contain many genes, and his study only showed a statistical tendency; it was by no means conclusive. But it did indicate a serious possibility that, one day, we may isolate the "gay gene" or, more likely, genes. This might reveal that homosexuality is 100 percent genetically determined or that it is 5, 10, or 70 percent determined, with the remainder due to the influences of the environment. Most geneticists would probably say there is likely to be a substantial genetic element—and that means we may soon be able to produce precise statistical forecasts of the likelihood of an unborn baby being gay. Parents would, therefore, have a choice.

This is how one geneticist responded: "There are probably few families that are overjoyed when they learn their child is not going to produce grandchildren. And people who are homosexual have a rougher time—you don't want your child to have a rougher time."

"What would having that choice do to society?" I asked.

"I don't know—make it duller, make a lot of mothers happier. I think most people would avoid having a homosexual child. And so it might be bad for society and good for individuals. There'd probably be less ballet."

The seriousness of the point is clear. However enlightened we may have become about homosexuality, *if offered the choice*, many, perhaps the overwhelming majority, would decide not to have a homosexual baby. At one level this means that a contemporary Michelangelo would stand little chance of being born. This may not strike the prospective mother as a serious concern; she is, after all, highly unlikely to be bearing a Michelangelo. But, at another level, it prevents the birth of any number of people whose place in and contribution to society may well be intimately connected to their sexuality. It would push society toward a preconceived idea of normality.

But the question then becomes, If you don't like this idea, what do you do about it? Peter Goodfellow, a British geneticist, observes that "Abortion figures across all the developed nations are roughly the same—95 to 97 percent are done for social reasons. In some eastern countries abortion is done because it's cheaper than birth control. We have to confront that. And, if we accept it, then I think we have to accept it when people decide they want to abort a fetus with a predisposition to homosexuality if that is their choice. I don't like it, but I can't see any other way to square these two components."

There is no way one can argue that widespread abortion of homosexually predisposed fetuses is anything other than eugenics. It is a clear program of manipulation of the gene pool and obviously a method of selecting for babies that are "well-born" insofar as people decide that to be born homosexual is not to be well-born. But it is not state-sponsored eugenics. Quite clearly, no Western democracy is currently likely to advocate or impose the abortion of homosexuals. Any such move would be regarded as intolerable. But is any such state likely to *prevent* the abortion of homosexuals? As Goodfellow points out, logic would be against any such attempt since abortions for all kinds of reasons are already permitted. In addition, many people would regard the attempt to prevent such abortions as an outrageous infringement of personal liberty; it would, in effect, be forcing some people to have gay children.

So what we face here is, in fact, the opposite of state-sponsored eugenics. This is the eugenics that happens when the state is specifically excluded from reproductive decisions. It is the eugenics of the free market, and it results inevitably from a combination of the current quasireligious faith in the absolute virtues of unfettered markets and the rapid growth of new genetic knowledge. The whole point is that we are about to be deluged with offers of choices. And each of those choices has the potential to produce fundamental conflicts between the public and the private realms. We may publicly agree that suppressing homosexuality is wrong, but privately we may decide we do not want a homosexual child. Let other people give birth to the gays to whom we will subsequently extend such tolerance. A few million such private decisions will constitute a very public program of eugenic suppression of homosexuality, a program that could well prove more effective than anything attempted by Hitler. The free market takes over where Nazism left off.

Homosexuality is, of course, just one example and is admittedly chosen to deliberately dramatize the point. But there are hundreds of other examples. At one end of the scale may be the abortion of fetuses suffering from devastating disorders like muscular dystrophy. At the other end may be the abortion of fetuses with brown rather than blue eyes. Where does homosexuality fall on that scale? And in time it will not be just a case of abortion. Babies tuned by genetic engineering or preselection of embryos—a technique already used to avoid embryos known to be genetically disordered—will be available to prospective parents. What will they do? And are we worried about what they will do just because, with Goodfellow, we "don't like it"? Or is there something more than mere distaste involved in this?

I believe the following: Children occupy a critical position on the frontier between the public and the private. For society as a whole they are the future, and all developed nations go to enormous lengths—notably through compulsory education—to ensure that their children are fit to inherit that future. When it comes to our children, we all accept some restrictions on our freedom; we are not free, for example, to deny our children education. So all children are, to some extent, public property in that the state regards itself as having a vital interest in some aspects of their upbringing.

On the other hand, of course, children are part of us as parents. I do not habitually look at my daughter and see her as the partial child of the British government, however much, objectively, I may know that to be the case. Rather, I look at her and see her as the closest, most precious thing to me. Now, in general, these two ways of looking at my daughter do not conflict. Both I and the British government want my daughter to be as successful as she possibly can be at whatever she chooses to do. I think this will be good for her; the government thinks it will be good for the nation as a whole. No problem.

But the government's desire for my daughter's success is only part of a generalized wish that all British children do well; mine is a very specific wish that my child does particularly well. And, because we live in a competitive world in which success involves doing better than others, that means I want her to be more successful than other British children. I can try and opt out of the overtones of brutality in this idea by saying I just want my child to be happy, or I can say that my desire for my child's betterment is good for society since, if everybody behaves that way, there will be a general pressure to produce a better world. There is nothing wrong with either of these arguments, but they do not in any way affect my principal point: that people in general are powerfully driven to gain a competitive advantage for their children or, at least, to ensure that they are not at a competitive disadvantage.

If offered the choice, therefore, they will find it very difficult, probably impossible, not to accept technological methods of improvement. But what would constitute improvement? Is it an improvement not to be homosexual? Or, to use another, less controversial example, is it an improvement not to be short? Not being short is, in fact, an improvement already available in the form of human growth hormone. But what is short? Parents whose son may grow up to be, for example, five feet four inches tall may say this is short and that he will be, as a result, disadvantaged. Research has shown that short people are not as respected as tall persons (Randy Newman has written a famous, cuttingly ironic song about this) and, consequently, they are less successful. So these parents want their son to be at least within a height range in which he will not be called "short." But, in fact, five feet four inches is not that short. It is simply at the shorter end of the variation in the population as a whole. And if everyone who is five feet four inches tall takes growth hormone, that variation will change. Maybe five feet eight inches will become noticeably short, and parents of sons who look as though they are going to be no more than that may demand growth hormone. Plus, if we take the absolutely libertarian view, what's the problem? Let anybody have growth hormone who wants it. The species will grow taller and then, maybe, being short will be prized. The free market will balance out the competing demands.

The central problem is, I think, that most people want the same things for their children—they want them to be smart, healthy, and generally normal. This is a problem in genetics, because insofar as we can choose such things, we will, and that will produce eugenic effects. Perhaps we will say that smarter and healthier are good eugenic effects, but even these can be ambiguous. Shortness may be seen as unhealthy, but is it? Being bad at math may be seen as stupid, but is it? Perhaps we need some people to be short and perhaps we need some people to be bad at math; apart from anything else, it may well mean they are good at history or some other worthwhile subject.

When we come to the idea of the normal, the whole issue becomes absurdly ambiguous. "Normal" is a safe strategy for our own child. Yet we often admire the abnormal—the socially difficult Mozart, the homosexual Michelangelo, the arrogant James Joyce. Indeed, when it comes to genius, only Shakespeare seems to rank anywhere near normal, and there is a strong theory that he was bisexual. But, when it comes to our own children, we prefer normal as the best, low-risk approach. It presents the fewest problems and the best chance of social integration. And, after all, it would scarcely make sense to cultivate so-cial delinquency on the off-chance that your child would turn out to be a Mozart.

Our children are perhaps the most important part of the whole aspi-rational mood of our society. We have come to regard "improvement" as a virtuous aspect of the consumer society in so many areas—from the state of our psyches to the state of our pectorals—that it is scarcely no-ticeable when we transfer the same idea to the bearing of children and thus slip into what is, in effect, a program of consumer eugenics. I can-not put this better than does the historian Mitchell G. Ash, who writes, "Does it really make a difference if eugenical *practices* are now taboo in government policy, if eugenical *discourse* is still very much alive in all our talk about 'quality of life' or even 'quality children'—internalized as part of the consumerist value system that unites us more than anything else. If the last question has made you uncomfortable, that was my in-tention." Mine too.

The general response from geneticists who regard all this as apoca-lyptic scaremongering is that the gene pool is so vast and so varied that microeugenics of this kind will not have any macro effects. Further-more, there are few significant traits about which it would be possible to make clear decisions. Francis Collins, head of the genome project in Bethesda, makes this point: "Suppose somebody comes out and says that if you have the B version of a gene, your IQ will on the average be three points higher. It probably will happen. Will that destabilize our society? We want kids with the B version, yuppies will say, but it's an ir-rational thing to do; intelligence is such a broad variant you would never notice the difference in an individual case."

"Spend more time with your kid," says geneticist Eric Lander. "That would make more difference."

The point both men are making is the reasonable one that, since our natures are made up of a complex mix of genetics and environment and since genetics is relatively difficult to alter, why not concentrate on the environment? For Collins, it is irrational to intervene eugenically be-cause, amidst all the environmental "noise," a few points on your child's IQ will make very little difference. For Lander, it is irrational because you can more easily help your child simply by taking greater care in his upbringing.

Others in the scientific community have not been so reluctant to confront the possibility of eugenic interventions. Daniel Koshland, a molecular biologist, asked in 1989, "If a child destined to have permanently low IQ could be cured by replacing a gene, would anyone really argue with that?" And he added: "It is a short step from that decision to improving a normal IQ. Is there an argument against making superior individuals? Not superior morally, and not superior philosophically, just superior in certain skills; better at computers, better as musicians, better physically. As society gets more complex, perhaps it must select for individuals more capable of coping with its complex problems. . . ."

Of course, "selecting for individuals" means a full-blooded eugenic program. This is unlikely to happen in the West, at least. State intervention in reproduction is and is likely to remain politically inconceivable. Singapore to some extent and China to a much greater extent have adopted eugenic policies—"Idiots give birth to idiots," was the *Chinese Peasants Daily*'s version of the basic genetic wisdom that like produces like—but no Western country would be so willing to sacrifice the personal in favor of the political. Perhaps, as Jeremy Rifkin has suggested, the multinational corporations will take control of the human gene pool. But it is more likely that the new eugenics will emerge quietly through the workings of the market and consumer society. "Once we have left the garden of genetic innocence," wrote the philosopher Philip Kitcher, "some form of eugenics is inescapable, and our first task must be to discover where among the available options we can find the safest home."

Kitcher's book, *The Lives to Come*, published in 1996, is perhaps the most thoughtful attempt to come to terms with this. For example, he addresses the problem of eugenic choices being available only to the rich: "If prenatal testing for genetic diseases is often used by members of more privileged strata of society and far more rarely by the underprivileged, then the genetic conditions the affluent are concerned to avoid will be far more common among the poor—they will become 'lower-class' diseases, other people's problems. Interest in finding methods of treatment or for providing supportive environments for those born with the diseases may well wane."

In addition, Kitcher sees that the eugenics of abortion amounts to a counsel of despair. We may choose to abort a Down's syndrome child against our better judgment simply because society's attitudes to such children are inadequate:

> Individual choices are not made in a social vacuum, and unless changes in social attitudes keep pace with the proliferation of genetic tests, we can anticipate that many future prospective parents, acting to avoid misery for potential children, will have to bow to social attitudes they reject and resent. They will have to choose abortion even though they believe that a more caring or less prejudiced society might have enabled the child who would have been born to lead a happy and fulfilling life. Laissez-faire eugenics is in danger of retaining the most disturbing aspect of its historical predecessors—the tendency to try to transform the population in a particular direction, not to avoid suffering but to reflect a set of social values.

Kitcher's solution is what he calls utopian eugenics:

> Utopian eugenics would use reliable genetic information in prenatal tests that would be available equally to all citizens. Although there would be widespread public discussion of values and of the social consequences of individual decisions, there would be no societally imposed restrictions on reproductive choices—citizens would be educated but not coerced. Finally, there would be universally shared respect for difference coupled with a public commitment to realizing the potential of all those who are born.

The problem, as Kitcher well knows, is that this is indeed *utopian* eugenics. His book makes it clear that handling the eugenic temptations offered by genetics will require supremely wise, ordered, and well-integrated societies. Otherwise the new eugenics may turn out to be as divisive and brutal as the old. That we do not have such wise societies is, effectively, the subtext of Kitcher's book. We have, therefore, to become much better people than we are. This is, to say the least, unlikely.

In 1989 a European Parliament committee report recommended that "each generation must be allowed to wrestle with human nature as it is given to them, and not with the irreversible biological results of their forebears' actions." This report was heavily influenced by both Roman Catholic and environmentalist thinking and it evokes a moral vision utterly opposed to the principles of eugenics, new and old. It says we cannot impose on future generations our conceptions of biological improvement because to do so represents an assault on human dignity. For it is the struggle with the givens of human nature that defines humanity, not the progressive effort to transform that nature.

Such idealism already seems almost quaint, for we have, as Kitcher says, left "the garden of genetic innocence" and entered a new world in which lives may be planned, controlled, or deleted by conscious human choices rather than merely lived. We may devise all sorts of euphemisms, but, one way or another, the future is eugenic. Michelangelo might never happen and foreheads might be tattooed. Who or what will we then be?

chapter 5

The Mighty Gene

The Mighty Gene

Innocent of what?
—SHERIFF LITTLE BILL, in the film *Unforgiven*

Behind the shifting face of personality is a hard nugget
of self, a genetic gift. . . . Biology is our hidden fate.
—CAMILLE PAGLIA

The secret is too plain. The pity of it smarts,
Makes hot tears spurt: that the soul is not a soul,
Has no secret, is small, and it fits
Its hollow perfectly. . . .
—JOHN ASHBERY

Genocentrism is the new word used to describe the belief that the gene is at the center of all things. It is now a widespread belief among both scientists and lay people. It has taken hold very rapidly. Before the nature of DNA was unraveled in the 1950s, the gene existed only in theory, not in fact. But, after Watson and Crick's discovery, it soon became clear that the gene was a length of DNA, a sequence of nucleotide bases that provided the code for the production of a protein. This was, as many have said, the Holy Grail of biology. It gave the subject a basic unit that seemed as hard and clear as the physicist's atom.

Once the news escaped from the laboratory, lay people became aware that the black box of life was finally being opened by the scientists. Soon the gene became all-powerful. Media coverage of advances in genetics intensified. The idea of a biological destiny fixed at conception took hold. It is now commonplace to hear people explaining some foible or other by saying, "It's in my genes." The phrase echoes the fatalism of the old saying, "It's in my stars," and it is significant that the idea of genes as destiny has taken on something of the popular force of astrology. DNA has become, in popular terms, magic. It seems to embody all identity, history, and meaning. Where once there were demons and witches, then neuroses and the traumas of childhood, now there are genes.

This has clear political implications. Before genocentrism, the prevailing view was that people were essentially equal. After genocentrism, the view becomes unsustainable. The genes reveal the profound differences with which we begin the race of life. Now, if we are to believe in human equality, we cannot do so on scientific grounds. We must devise a metaphysical basis that precedes the insights of science.

Yet a scientific war has been fought over the truth or otherwise of genocentrism. On one side writers like Stephen Jay Gould have argued that it is fundamentally false, that the genocentrists have oversimplified their case. On the other side, writers like Richard Dawkins have eloquently shown the awesome explanatory power of the view that it is the "selfish gene" that drives all life and, therefore, all human affairs. There are also writers like Harvard biologist E. O. Wilson who take the broader cultural view that the realization of our biological destiny has become the central imperative of human life. At last evolution and the mighty gene give us a scientific basis for the moral and spiritual life of humanity.

The details of these scientific arguments are not important here. All that matters is that both in science and in the rest of the world, genocentrism is now the dominant orthodoxy. Gould and his supporters, right or wrong, are in a minority and the popular force of the image of the mighty, controlling gene seems unstoppable. My task is to look at the human implications of the triumph of the genocentrists.

The British geneticist Peter Goodfellow is one of the most thoughtful of scientists. He has the exciting and, among scientists, rare habit of setting himself broad moral and social problems to which he knows he has no answer. So, in the summer of 1996 in a park near his home in Hampstead, London, we talked about the exact meaning of the relationship between genes and behavior, and he offered this vivid—and, I should add, entirely imaginary—illustration of the issue: "Say we've just discovered the gene which identifies a predisposition as to whether you will be a transvestite or not. This means that those individuals who carry this gene are much more likely to dress up in women's clothing. Unfortunately, although I wear a dress every Thursday at the lab, I

THE MIGHTY GENE · 93

don't carry this gene. Therefore I must be perverse. Where have we got from knowing this piece of information? The real issue is people's response to you if you wear women's clothes."

This is a profound point. Forget for the moment whether you personally regard transvestism as perverse, distasteful, and wrong or a legitimate expression of sexuality; merely accept that, in the real world, it is a habit that provokes both responses. To some people a transvestite is guilty of something, to others he is not. Now if we find a gene that either makes him a transvestite or, at least, encourages him to be one, does that make him less "guilty"?

As the philosopher Dan W. Brock has put it: "If a person's genetic structure is a principal cause of behavior and that genetic structure is completely beyond the individual's control, can an individual justifiably be held responsible for the resultant behavior?"

Logically he is less responsible—he has, after all, made no conscious decision to have that gene; it was bequeathed to him by his parents. Therefore, by the same logic, if he does not have that gene, he must be more responsible if he dresses up in women's clothes. Even people who take the most liberal view of transvestism must be driven to make some sort of distinction between a condition that is genetically determined and one that, in some sense, appears to be chosen. After all, having naturally red hair means one thing; dying your hair red means another.

To take the point even further, is the genetic transvestite really less responsible? If we hold the view that the genes define what we are, then this transvestite's gene is an integral part of what he is. And, since we judge people by what they are, then we must, ultimately, be judging a person's genes. Even the man who does not have the transvestite gene and yet who dresses up as a woman might have another collection of genes that makes him prone, for quite different genetic reasons, to this kind of aberrant behavior—say, genes that fill him with the desire to shock.

At this point such arguments tend to be made even more complex by the addition of the environmental factor. Psychoanalytic explanations of transvestism would focus on the man's upbringing, specifically his childhood and the relations within his family. The man who dresses in women's clothes without possessing the gene would be seen as an environmentally determined transvestite. And even the geneticist who had identified a transvestite as having the gene would be likely to accept an environmental element in his condition. After all, he would be obliged to admit, not everybody with the gene is a transvestite.

The convolutions of this point are an expression of the sort of complications that arise when we start to consider the effects of the mighty gene in the real world. The rest of this chapter is an attempt to assess and catalogue those effects, though, in truth, much of what I have to say is implied by Goodfellow's thought experiment.

One important point, however, is not present in that quotation. Because of the way statistics and genetics work together, it is routinely said that some human trait is x per cent inherited and y per cent environmental. What this means is that statistical methods have shown that in a certain number of cases a trait is passed down through the generations. In the cases where it occurs without any sign of being inherited, or, conversely, where it is not passed down, or appears to be suppressed, there must be other factors—environmental factors—at work.

This seems reasonable enough, except that it does not—cannot—mean what it appears to mean and what it is often taken to mean, even by the scientists who advance the idea. It does not mean that the environmental element is nongenetic. The reason is that when we talk of the human environment, we mean, overwhelmingly, other people. This is not just the other people with whom we come into direct contact; it is also the people who build our cars and houses, write the books and newspapers we read, and run the companies for which we work. It also means the millions of people, mostly dead, who have created our language and culture. All these people have genes and environments, but their environments will also largely be other people. In short, the human environment is, in reality, no more than *other people's genes.* There is no mysterious, external factor embodied in the word "environment," no escape from the might of the gene. This is the logic of the genocentric view of the world.

That said, let me start with the underlying theme of Goodfellow's thought experiment: guilt. In the eighties in Georgia a young woman was given a life sentence for murdering her mother. In mitigation she had pleaded that, as she was at risk of Huntington's disease, the murder may have been the first sign of onset. Since her father had died of the disease and since the Huntington's gene is dominant, she did, indeed, have a 50 percent chance of inheriting the condition. The judge did not accept the argument. But when, a few years later, she contracted the disease, he recalled the case and released her. The judge had decided that her guilt was removed by the gene.

In 1991 Stephen Mobley killed John Collins while robbing his pizza store near Atlanta. It was a clear-cut case and Mobley was sentenced to death. There appeared to be no mitigating circumstances to which his lawyers could appeal. But when they began researching his family history, they found four generations of crime, violence, and antisocial behavior. So bad was the family reputation that local wisdom said that the best thing about being a Mobley was that you didn't have to marry one. The lawyers appealed the death sentence on the basis that Mobley had inherited a defect which could, conceivably, be corrected. Therefore not only was he less guilty since he was acting on an impulse of his genes, but he was also capable, in theory, of being "cured."

This "criminal gene" case received worldwide publicity for an idea that had, in fact, originally been born in the Netherlands. There, research into a family in which half the males were angry, antisocial, and possessed low IQs had shown a pattern of enzyme peculiarities. Specifically, they had a mutation in the gene which coded for monoamine oxidase, a so-called "neurotransmitter" which metabolizes adrenaline, the hormone that pumps up the body to meet a threat either by running away or fighting. The defect was on the X chromosome, so it would tend to be carried by women and expressed in men. The men in this family were not metabolizing adrenaline properly, so they would find themselves in a "fight or flight" state more often than anybody else. They would permanently be aggressive and ready for action. This research formed the basis of the Mobley defense.

Clearly there have always been many types of arguments for mitigation in criminal cases—I did it because I was giving up smoking, because I was abused as a child, because of temporary insanity, and so on. But none is as fundamental as the "criminal gene" argument, for it raises the basic question: What exactly is on trial? Is it Stephen Mobley, his genes, or the act itself?

To ask this question is to expose a kind of legal fiction on which our system of justice has always relied. As Deborah Denno, professor of law at Fordham University in New York City, put it to me: "We treat people as being autonomous and willed human beings, not because they are, but because that's the way the criminal law treats people. There has never been an assumption that we really are autonomous beings; it is simply the most effective method for a criminal justice system. How else are we to do it? Saying we are automata would excuse anything."

If we are automata, or robots run for the benefit of the genes, then potentially any crime can be explained away in entirely guilt-free terms. Denno's observation goes to the heart of the impact of scientific knowledge on our moral world. Science tends to create explanatory systems which do not respect stable categories like "the individual." If science asks "Why?" of a crime like murder, it cannot be content with the answer: Because he is a bad man. The reason is that this is not an answer within the explanatory framework of science, most obviously because the word "bad" cannot be scientifically validated. Instead, science comes up with explanations that do not depend on such words. But, in doing so, it tends to abolish the willed, autonomous individual. If science says that Mobley committed this murder because of a genetic predisposition, then the implication is that Mobley is, indeed, an automaton and not, ultimately, guilty.

As Dan W. Brock points out, there is an intrinsic tension between morality and the scientific view of human behavior. Science treats human behavior as just one more natural phenomenon. But that is to treat a human act as simply the end of a causal chain rather than as a morally significant event. Genetics aspires to read the causal chain.

"Our fundamental conception of ourselves as persons," Brock has written, "and, more specifically, our belief in ourselves as beings with sufficient free will to be responsible agents will come into question."

If we are all biologically controlled automata, then we are subject to forces beyond our control that are, somehow, *inside* us. In fact, it becomes difficult to draw the line between ourselves and these forces. Mobley's genes made him do it, but his genes are him. It is difficult to see, once we start going down such a path, where we could stop short of a total dissolution of the concept of a morally coherent and responsible individual.

Of course, we could patch up this state of affairs. We could put an artificial limit on the weight attached to genetic explanations. Or we could decide that this individual with this predisposition did commit this crime, whereas this other individual, who happens to have the same genes, did not. In the differences between these two individuals lies a space large enough to be called moral responsibility—one makes the moral choice to thwart his genes, the other does not. But then environmental factors could intervene—perhaps the man who does not commit the crime had a better upbringing. This could again mean he was not morally better, merely luckier. And, anyway, as I have said, the environment is just other people's genes. Either way we end up as robots.

Perhaps, in the end, the traditional idea of guilt and innocence will simply evaporate to be replaced by the purely practical idea of locking people up who are dangerous. But then why not lock them up *before* they do anything on the basis of their "bad" genes? It would be logical and, in the absence of any moral concept of praise or blame, an entirely virtuous thing to do. Indeed, people might start to demand we do it. I can just imagine the concerned TV news reporters cornering some hapless genticist: "You mean you knew he *might* do this and you did nothing?" Against such pressure, it would be difficult to argue that he was not locked up because there was a chance he might not do it. We are a society that, in all areas, is obsessed with reducing every possible risk.

And then, since we all almost certainly possess genes that predispose us to do something wrong, what becomes of the possibility of innocence? It becomes a concept purely related to one particular incident rather than a general condition. As Gene Hackman's savage sheriff Little Bill says in the film *Unforgiven*: "Innocent of what?"

Ideas of guilt and innocence are fundamental to the way we conduct ourselves, not just in the courts but also in our daily lives. It is impossible to live and interact with others without having a sense of individuals as responsible agents, subject to praise and blame. This is just the way the language and the culture work. There may not be any other way they *could* work. Genetics puts a new kind of pressure on the formal and informal systems of justice through which we live. It does so by calling into question the meaning and reality both of the self and of the moral conviction involved in any act of praising or blaming. In the courts and in our lives we may be able to deal with this pressure through any number of subtle compromises. But will we be able to do that in society as a whole?

The issue of criminal justice leads directly to the wider issue of social justice. The idea of equality at the starting line—equality of opportunity—has always been built into a certain progressive, liberal view of the world that wishes to believe in the overwhelming importance of the environment in determining human character. Once the environment was equalized, then equality of opportunity should, logically, lead to equality of outcome. Liberally inclined geneticists—and this includes almost all those I have met—know that the latter part of this process is too much to hope for. There are fundamental, genetically determined differences between people which will inevitably mean that equality of outcome cannot be guaranteed.

Yet they wish to stick to a broadly liberal ideal of equality. They invariably do so by adopting a form of argument which is, in effect, an extension of the Darwinian insight of the connectedness of all life. They argue that the revelations of genetics are not about the depth of our differences, but about the extent of our similarities. We share DNA or RNA with every other living creature and, with each other, we share almost the entire nucleotide sequence. At the molecular level we are staggeringly similar, remarkably equal. This, they say, is one of the key insights of modern genetics—that, beneath the skin, we are all almost exactly the same. If DNA is the modern version of the soul, then it is a soul we all share and the message coming from this soul is optimistic: We are all one. Genetics and genocentrism are good for us.

So persuasive is this message that even antigenocentrists like Dorothy Nelkin and M. Susan Lindee embrace it—without, apparently, noticing the inconsistency it creates in their position:

> It is especially ironic that DNA has become a cultural resource for the construction of differences, for one of the insights of contemporary genomics research is the profound similarity, at the level of the DNA, among human beings and, indeed, between humans and other species. We differ from the chimpanzee by only one base pair out of a hundred—

1 percent—and from each other by less than 0.1 percent. The cultural lesson of the Human Genome Project could be that we are all very much alike, but instead contemporary molecular genetics has been folded into enduring debates about group inferiority. Scientists have participated in these debates by seeking genes for homosexuality and alcoholism, genes for caring and genes for criminality. This research and the ideological narratives that undergird it have significant social meaning and policy implications.

The inconsistency arises from the way Nelkin and Lindee say the genes are unimportant when they reveal an undesired inequality, but important when they reveal a desired equality. You cannot have it both ways. But, in general, that is the argument of those who are optimistic about the social impact of genetics. It is an argument that is as consoling as a Coke commercial and it is, on closer inspection, if not exactly wrong, then at least remarkably feeble.

This is how the ethicist Marc A. Lappé puts the opposing argument: "The products of the genome project may throw into stark relief the paradox of a society based on the premise of equal standing at creation and one that is found to be composed of a genetically heterogeneous group of subpopulations with qualitatively different frequencies of heritable traits."

The optimists may be right in crude numerical terms when they say that our molecular differences from each other are small. But they are wrong to think this insight will carry any persuasive social weight. For no matter how often you tell people they are only 1 percent different from a chimpanzee, they will continue stubbornly to believe they are very different indeed. And, if you point out they are only 0.1 percent different from Stephen Mobley, they will prove even more stubborn in their insistence that they are nothing like the man.

And they will be right. For it is unarguably true that the differences between a monkey and a human are huge and the differences between a brutal killer and one who would never kill are also huge. Anybody who denies that they are huge is quite simply excluding themselves from serious debate. The point is—as every geneticist, however liberal-minded, knows perfectly well—that a small number of nucleotides can make a very big difference. One wrong base pair can give you sickle-cell disease; one malfunctioning gene can, apparently, make you a killer. The fact that, at the molecular level, the difference appears small is irrelevant because, at the molecular level, *everything* appears small. And, besides, the whole of modern science from quantum theory to chaos theory has successfully persuaded us of the fact that small things make big differences.

The truth is that, far from asserting equality, genetics does the reverse. It affirms inequality by giving it a simple, material reality. If, instead of producing vague and approximate theories about Stephen Mobley's background, you can point to a specific point on his X chromosome, maybe even to a specific base pair, then you will convince people not of his slight difference from the normal, but of his *ultimate* difference.

"What is clear," continues Lappé, "is that even a partial picture of the genetic landscape that defines the molecular differences among human individuals will reveal more in its nonuniformity than in its hoped for universality."

I cannot overstate the importance of this point. The molecular affirmation of human difference may be the most important and—for good or ill—persuasive insight of contemporary genetics. It has implications for the treatment of disease and it may come to define social policy as well as social attitudes. It puts yet more pressure on the Christian and Enlightenment ideals of the moral absolute of the individual, and, most spectacularly, it produces a grand historical irony.

For, with the ending of the Cold War, the Western democratic system is globally triumphant. That system is based on equality and freedom of inquiry; it is driven by science and technology. And yet here is a scientific insight, one that will soon have radical technological applications, that threatens to undermine equality by saying that, at the very core of their being, human beings are unequal. One half of the Western system—the scientific—has risen up to threaten the existence of the other—the democratic. Preserving an ideal of equality against the genetic insight would require a metaphysical effort of superhuman proportions, and we are neither superhuman nor, thanks to the persuasive powers of science, metaphysically inclined.

Much of the effect of this point lies in the future. The impact on the judicial system is just the start. In due course it will affect politics, both domestic and international. But I can illustrate its direct effects by using a recent example of a violent and embittered controversy that arose directly from a book that affirmed fundamental genetic differences between races: *The Bell Curve: Intelligence and Class Structure in American Life*, by Richard J. Herrnstein and Charles Murray, published in 1994.

The book needs to be handled with care and to be placed very precisely in a certain context. The testing of intelligence has been an important preoccupation of those seeking to find heritable traits in human beings. Ever since the IQ tests conducted on American soldiers in the First World War, there have been large-scale attempts to establish intelligence trends in populations employing increasingly refined methods of measurement. Intelligence has always seemed to be, perhaps understandably, the human characteristic that obsesses us the most and, as a result, it is the one we aspire most energetically to measure.

Yet, oddly enough, we seldom want to hear the results. Perhaps it is simply because the news tends to be bad. Mass testing has tended to show, for one reason or another, either that IQ is in general decline or that some group is particularly bright or stupid. Even when it is the IQ of a single person that is involved, people can become queasy, especially if it is their child. There is something awfully definitive about that single figure and something that feels wrong about reducing something as complex as intelligence to one number.

In 1982, Stephen Jay Gould tracked the whole history of intelligence testing in a book whose title states his conclusion: *The Mismeasure of Man*. This was a brilliant polemic which took that queasy feeling and made it respectable. Gould's argument was that intelligence testing was just a new twist to an old story that began with a device, dreamed up by Plato, that would ensure a stable society by telling people a lie—that they were natural born commanders, auxiliaries, or craftsmen. By telling them there was something built into their natures that made them especially suited to their job, Plato thought he could avoid conflict and social disruption and guarantee order. People would always feel they were fulfilling their destiny

"The same tale, in different versions," wrote Gould, "has been promulgated and believed ever since. The justification for ranking groups by inborn worth has varied with the tides of Western history. Plato relied upon dialectic, the Church upon dogma. For the past two centuries, scientific claims have become the primary agent for validating Plato's myth."

And Gould adds, significantly: ". . . I was inspired to write this book because biological determinism is rising in popularity again, as it always does in times of political retrenchment. The cocktail party circuit has been buzzing with its usual profundity about innate aggression, sex roles, and the naked ape. Millions of people are now suspecting that their social prejudices are scientific facts after all."

Gould was rebelling against the cultural change I described in the last chapter from a culturally deterministic to a genetically deterministic view of the world. His case was that the history of intelligence testing was a history of bad science—a mad cascade of theories and experiments, all of which were hopelessly compromised by the prejudices and presuppositions of the theorists and experimenters. Usually they were either implicit or explicit sexists or racists, convinced that women, blacks, or Native Americans were innately inferior, or they were so committed to the idea that there was a genetic basis of intelligence that they could not see how wrong they were.

In spite of the rising biological determinism that Gould had noted, this book effectively became the popular orthodoxy for at least the next ten years and, at most cocktail parties, it probably still is. Gould is a persuasive writer, and his central point—that science is always the servant of a particular culture—is both true and, in application, humane.

"We pass through this world but once," he writes. "Few tragedies can be more extensive than the stunting of life, few injustices deeper than the denial of an opportunity to strive or even to hope, by a limit imposed from without, but falsely identified as lying within."

Unfortunately, Gould's idealism was based on a belief that was plainly unsustainable. Certainly he may have successfully discredited a great deal of research into IQ and, equally certainly, he may have raised serious objections to the means of measurement. But he did so in the apparent belief that IQ was not definably heritable and neither was it measurable—at least not by any method that would reduce it to a simple number. This is only a belief and it may easily be proved false.

THE MIGHTY GENE · 103

We may have got it wrong in the past, but that was no reason to believe we could not get it right in the future. With better methods, perhaps we could show heritability of IQ and find a means of more accurate measurement. There was nothing in Gould's argument that said this was impossible, merely that it hadn't happened yet.

(Before moving on to the next stage of this argument, I need to issue a disclaimer. I have issued this disclaimer in various forms elsewhere in the book, but here, for reasons that will become clear, it needs to be written on stone in letters of fire: Precisely what I personally think about all of this will become apparent later. For the moment I am dealing with arguments from particular perspectives in order to draw out their full significance. I do not necessarily believe any of these arguments are true; if the whole gene-centered view of things turns out to be false, then many of these arguments will also be false. But there is a logic involved which has to be followed through if we are ever to understand what genetics is really all about.)

Hidden beneath the anti-IQ orthodoxy which Gould created, there remained a widespread belief among scientists in various disciplines both that there was, indeed, a large and probably decisive inherited component in intelligence, and that there was some objectively measurable human quality defined by the word "intelligence." In ordinary experience neither belief is particularly startling. Intelligent people do appear, on the whole, to have intelligent children, and we do have some possibly vague but nonetheless real conception of intelligence. Both are commonsense insights which, for a large number of people, were successfully suppressed by Gould's brilliance.

The Bell Curve came along as a sharp corrective to this state of affairs. The book is plainly, at one level, a response to Gould and the prevailing, liberal orthodoxy. It is a response made for hard, economic reasons. The authors believe it is becoming impractical to deny the obvious in order to salve liberal consciences; it is simply becoming too expensive. Murray and Herrnstein cite the 1971 U.S. Supreme Court decision of Griggs *v.* Duke Power Company. Since then, they say, "no large employer has been able to hire from the top down based on intelligence tests." It was a decision that, according to various estimates, cost the U.S. economy between $13 billion and $80 billion in 1980. Further-more, because employers are prevented from testing applicants using any method that may be racially discriminatory, anything resembling standard IQ tests, at which blacks have invariably scored worse than whites, cannot be used. So the best people are not being chosen. Meanwhile, radical and often concealed positive discrimination policies have resulted in damaging distortions in the higher education system that have increased, not reduced, racial divisions. In other words, it is all very well to believe people are biologically equal, but when you start imposing that belief on the world, there is a high price to be paid.

Two further points form the basis of this book. First, the removal of various forms of discrimination—based on class or race—has massively reduced the environmental barriers to success. The barriers that remain must, therefore, be biological.

"Putting it all together," Murray and Herrnstein wrote, "success and failure in the American economy, and all that goes with it, are increasingly a matter of the genes that people inherit."

Second, modern societies are very efficient when it comes to finding and rewarding intelligence. High IQ is the supreme quality required in the modern world, and all the democracies, but especially the United States, have found increasingly sophisticated ways of selecting the most intelligent from all areas of society and sequestering them within specific elite communities, academic or professional.

Many do not seem to understand the significance of this point. For example, when I spoke to the Johns Hopkins University geneticist Victor McKusick, he happened to have a copy of *The Bell Curve* in his bag. He said it was a "very nasty book as far as its conclusions were concerned," but added that "much of it is true." Then he went on to say that "there are many valuable qualities that are not measured by the IQ test." Murray and Herrnstein would certainly agree with this. But their point is that the way society is now organized means that what is measured by the IQ test is *overwhelmingly the most valuable quality*, determining material rewards and status far more consistently and for greater numbers of people than, say, sporting or artistic talent. In short, our liberal consciences may say IQ is not the only thing that matters, but our liberal societies behave as though it is.

The full significance of *The Bell Curve* can only be understood if this point is grasped. For, among other things, the book is a warning about the dangers of "cognitive stratification"—the division of society into levels of intelligence. Such a society will be the least plural ever devised; there will be simply no contact between the haves and the have-nots.

"Try to envision what will happen when 10 or 20 percent of the population has enough income to bypass the social institutions they don't like in ways that only the top 1 percent used to be able to do," say the authors. "Robert Reich has called it the 'secession of the successful.' The current symbol of this phenomenon is the gated community, secure behind its walls and guard posts."

Murray and Herrnstein say that blacks are, on average, significantly less intelligent than whites, and whites somewhat less intelligent than Asians. They also revive the old anxiety about dysgenesis—"Mounting evidence indicates that demographic trends are exerting downward pressure on the distribution of cognitive ability in the United States and that the pressures are strong enough to have social consequences." And they suggest that the present levels of crime and dysfunction within society may be a product of dysgenic pressures on the U.S. throughout this century. They also consistently make it clear that they believe IQ does seem to relate to morality. Low-IQ people tend to be worse people: "Perhaps the ethical principles for not committing crimes are less accessible (or less persuasive) to people of low intelligence. They find it harder to understand why robbing someone is wrong, find it harder to appreciate the values of civil and cooperative social life, and are accordingly less inhibited from acting in ways that are hurtful to other people and to the community at large."

Their conclusion is that we must, above all, be realistic: "Inequality of endowments, including intelligence, is a reality. Trying to pretend that inequality does not really exist has led to disaster. Trying to eradicate inequality with artificially manufactured outcomes has led to disaster. It is time for America once again to try living with inequality, as life is lived: understanding that each human being has strengths and weaknesses, qualities we admire and we do not admire, competencies and incompetencies, assets and debits; that the success of each human life is not measured externally but internally; that of all the rewards we can confer on each other, the most precious is a place as a valued fellow citizen."

This book, unsurprisingly, caused a storm, primarily because of its insistence on the IQ differences between races, but also because of its acceptance of the numerical reality and heritability of intelligence. For the purposes of this chapter, its importance lies not so much in its specific assertions as in the shape of its ideas and in the resistance to those ideas.

First, the resistance. One specific aspect of this is important. Murray and Herrnstein's approach is "top-down"—it studies patterns of inheritance by looking as their impact on large populations, and then reasons downward from this to what must be true at the level of the genes. One specific theme of the response to their ideas is that the bottom-uppers—the molecular geneticists—tend to be violently opposed to the book. Their argument is usually the one I mentioned earlier—that the study of genetics shows more similarities than dissimilarities between people. But, in the case of David Botstein at Stanford, the argument was primarily statistical. When I asked him about *The Bell Curve*, he leaped to his feet and started drawing graphs.

He was illustrating a thought experiment. Say, he explained, that we take 100 nine-year-old children from a town and ask each of them to run a hundred-yard sprint. We time them and we plot their performances on a graph. The distribution of times will show a small bell curve at the faster end of the scale—this will be those children that have previously trained—and a large bell curve covering the performance of the rest. Say we train all of these 100 children for a year and go through the same procedure again. This time we will have a single bell curve that will show a markedly better performance by all the children. But the point is that—and Botstein grew passionate when he reached the climax of his demonstration—there is no way we could predict the position of an individual on the second graph from his or her position on the first. And the reason we couldn't is that there are too many factors—too much noise—involved. There would be no way we could ever say that this aspect of an individual's ability is genetically determined, even with something as apparently simple as athletic ability. With intelligence, the difficulties would be completely unthinkable.

In fact, as Botstein probably knew, this is not an argument against Murray and Herrnstein. All he was saying is that you cannot leap from the statistics to an individual, and that is a point the authors themselves repeatedly make throughout *The Bell Curve*. They are concerned with mass tendencies, not with using those tendencies to judge or make predictions about any one individual. They know perfectly well that such judgments and forecasts are impossible.

But the fact that two top-downers and one bottom-upper confront each other in this way is important. We will see why when we look at the overall shape of Murray and Herrnstein's ideas.

In essence, *The Bell Curve* is a statement that the human sciences—so long patronized as "soft" in contrast to "hard" sciences like physics—can, in fact, be hard. For the authors, IQ is, indeed, a hard fact, and their evidence is the mountain of statistical correlation they provide. Although a hard scientist may be able to dismiss this or that correlation between, say, IQ scores and behavioral problems or poverty, he will find it more difficult to dismiss the huge range of correlations Murray and Herrnstein provide. And all of these point to some hard entity, some irreducible "truth," which is present within the results of IQ tests.

The book's further statement is that this truth—that there is a variation of this absolute quality called intelligence from individual to individual—is genetic. Not only are modern IQ tests carefully structured to eliminate cultural bias, but modern society has eliminated more and more of the social conditions that have tended to eliminate inequalities other than those that arise from the biological advantages or disadvantages of the individual at the starting line. So, for example, if everybody goes to school, then nobody will be disadvantaged because they don't go to school, and any differences in outcome will be more likely to be due to inborn qualities.

The point about these arguments is that there is nothing *necessarily* wrong with them. Indeed, it might be said that, from their top-down perspective, the authors are doing no more than pursuing the logic of the bottom-up molecular geneticists. Like the geneticists, Murray and Herrnstein also believe in the centrality of the genes. Certainly they may be wrong in detail and certainly their statistics may one day be shown to be ill-founded, but what they are saying is perfectly in line with the wisdom of the genocentric age in which we live. Murray and Herrnstein are as much expressions of the cult of the mighty gene as are Stephen Mobley's lawyers.

The trouble is that what they are saying is unpalatable. If this is the genetic truth, then it is a truth that divides. Their conclusion—that we must relearn the ability to internalize our sense of worth—is, they must know, utopian. The continuing reality of our societies is, as they correctly diagnose, that they are becoming more not less divided. That division, as, again, they correctly point out, is based on the creation of new intellectual elites that are based on the withdrawing of certain types of gifted people—they say they are simply those with the highest IQs—from mainstream society and their isolation in communities that are, increasingly, becoming accustomed to the bypassing of socially cohesive institutions. They hire their own guards rather than support the police. They buy private education rather than improve public schools. And all of this because, as Murray and Herrnstein see it, society has become so much more efficient at selecting for intelligence, at realizing the inborn biological basis of a certain form of excellence.

I have used *The Bell Curve* as a way of showing the sorts of issues that might arise from the deep perceptions of human difference that are intrinsic to genetics. Murray and Herrnstein are not molecular geneticists, and the sorts of points they make could not yet be made by molecular geneticists. But they could soon, and, indeed, the entire logic of their position is based on the hard truth of heritability that is also at the center of molecular genetics. And consider one claim of theirs in particular—that there is a link between moral behavior and IQ. This is truly a devastating idea in the context of traditional concepts of equality, for it encourages the view that there is no inner person who can be judged good or bad, that there is only this victim of inheritance whose goodness or badness is dependent not on some inner light but on his intellectual gifts. And that would be one further nail in the coffin of the Christian or Enlightenment soul, equal in the sight of God or man.

When confronted with Murray and Herrnstein, the big question is not whether they are wrong or right, because what they are saying is logically bound to be the outcome of the genetic enterprise. Rather the question is: Can we cope?

There is, however, an alternative, optimistic view of the way the mighty gene might function at the center of our society. E. O. Wilson created the discipline of sociobiology and applied it to humanity. Wilson believes that the revelation of our genetic heritage provides the foundation for a new, unifying myth—a myth that happens, as he says, to be true. Wilson's sociobiology as applied to humans proved unacceptable to many on the political left. It was felt to be another establishment attempt to preserve the status quo. Richard Lewontin, Steven Rose, and Leon J. Kamin damned the idea most eloquently in their 1984 book *Not in Our Genes*. "Over the past decade and a half," they write, "we have watched with concern the rising tide of biological determinist writing, with its increasingly grandiose claims to be able to locate the causes of the unequalities of status, wealth and power between classes, genders, and races in Western society in a reductionist theory of human nature."

And, specifically of Wilson's idea, they say: "Sociobiology is yet another attempt to put a natural scientific foundation under Adam Smith. It combines vulgar Mendelianism, vulgar Darwinism, and vulgar reductionism in the service of the status quo."

This form of argument exactly parallels Stephen Jay Gould's assault on intelligence testing and research into the heritability of intelligence. Like Gould, these authors pour extremely well-written scorn on the idea that human behavior can be read, as it were, through the genes, arguing instead for a more humane, less deterministic view which allows for the complexities created by human consciousness. Like Gould, they cannot necessarily be said to have made their case; even if they are right, they show only that the effort has proved wrong *so far*. And, again like Gould, they must be watching with dismay as, in recent years, there has been a resurgence in the very idea they were trying to discredit— the idea of biology rather than environment or culture as the determinant of our fate.

The Bell Curve has not been the only argument in favor of biology. In the field of anthropology there has also been a massive change of emphasis. This centers on the reputation and authority of one book: Margaret Mead's *Coming of Age in Samoa*. Mead wrote this book in the 1920s. It was based on her fieldwork among the islanders of Samoa. It portrayed an idyllic society in which the traumas of adolescence, so common in Western societies, simply did not happen. Teenagers freely made love and avoided the rebellions and repressions of industrialized societies. The message of the book was that culture, not biology, determined human behavior. And that message became the dominant anthropological orthodoxy. *Coming of Age* was regarded as a master-piece, a classic of scientific anthropology.

But, in 1983, the Australian anthropologist Derek Freeman published *Margaret Mead and Samoa*, a book that is now entitled *Margaret Mead and the Heretic*, and started the most violent controversy in the history of anthropology. Freeman set Mead's work in a particular context—a context in which there was a profound intellectual war going on between biological and cultural determinists. I have described one side of this war—that of the biologically deterministic eugenicists. The other side comprised those who were revolted by this orthodoxy and wanted to prove the centrality of culture as opposed to genes. Mead was on this side and she went to Samoa determined to prove her point.

Freeman says that, largely as a result of this bias, her observations were hopelessly flawed. She did not get close enough to the Samoans and she was the victim of a series of hoaxes by the young girls she claimed were so sexually liberated. Samoan society turns out to be as, if not more, stressful, repressive, and violent as any other. The sexual liberation she saw was in her imagination; in reality the Samoans adhere to a strict cult of virginity. And so on. Mead's status—which had been as one of the prophets of the belief that culture, not genetics, was the key determinant of human life—was massively compromised. The myth of the *tabula rasa*—the behavioral blank slate of the baby at birth—had been destroyed.

"We have reached a point," wrote Freeman, "at which the discipline of anthropology, if it is not to become isolated in a conceptual cul de sac, must abandon the paradigm . . . and must give full cognition to biology, as well as to culture, in the explanation of human behavior and institutions."

The further pro-gene development in recent years has been the resurrection of sociobiology under a new name—evolutionary psychology—and with a new confidence. Several factors have converged to make this happen: Wilson himself; the selfish gene view which has proved immensely successful as an explanatory system, at least within the confines of animal behavior; the increasing power of evolutionary thinking within nonbiological disciplines like sociology and psychology; and, finally, some extraordinary work in the application to evolution of game theory, a branch of mathematics created by the Hungarian mathematical genius John von Neumann in 1944 and introduced into biology by John Maynard Smith in the early seventies.

Game theory is the clearest way into the new sociobiology which aims to place the gene at the center of our understanding of human life. Game theory seeks to make sense of competition by analyzing different moves in as clear a mathematical way as possible. It is based on the insight that a single active agent in the world—say me, if I am the only human—can simply maximize his environment for his own benefit. But as soon as, say, you appear in my world, we will inevitably come into some form of conflict: I want to eat all the coconuts on this tree, but you want some as well, so we must evolve some strategy that may involve either cooperation or one of us killing the other. With the proliferation of agents, complexities mount.

Biology comes in when we realize that the agent does not have to be human. It can be a plant or an animal. This is exactly what evolution says, that organisms will compete against each other as in a game. While a plant or animal may not make rational "moves," it will definitely do things, such as attempt to grow in this particular spot or eat this particular food, which amount to moves and which will result in either success or failure. The living world, self-conscious or not, is full of players.

This way of studying evolution proves extraordinarily fertile because it allows scientists to assess complex systems in relatively simple terms and make testable forecasts about animal and, as we shall see, human behavior. Take the most famous game in game theory—Prisoner's Dilemma. You and a friend are both in prison awaiting trial on a false charge. The prosecutor makes you both the same offer. If neither of you confesses or implicates the other, you both get a short sentence. If you confess and he doesn't, you go free and he receives a life sentence. If you both confess you both get medium sentences. And, obviously, if you say nothing and your friend confesses, then you get life and he goes free.

The beauty of this game is twofold. First, it leads to incredible subtleties of calculation and, second, it represents a remarkable distillation of all the complexities in any confrontation, from the Cuban Missile Crisis to the sharing of a zebra corpse among a pride of lions. Millions of words have been written on Prisoner's Dilemma, so, instead of repeating them, I shall rely on the reader to run through a few of the possible strategies and their long-term implications. What at once becomes clear is that one theme dominates all others: the choice between niceness and nastiness.

The importance of this is that it offers a way of addressing one of the great problems of evolutionary theory, especially in its selfish gene form—the problem of altruism. If genes are merely blindly pursuing their own self-interest and all organisms are no more than the vehicles for that self-interest, why is anything—plant, animal, or human—ever nice to anything else? One theory used to be that there was something called "group selection," in which organisms cooperated for the greater good of the group as a whole. But this makes no sense in the context of the selfish gene because all the gene can possibly see is the survival of its own particular organism. This has been modified slightly to allow "kin selection"; since close relations share genes, it may well be in the interest of the gene to prompt its organism to help closely related others. If this particular gene cannot survive through this particular organism, then a precise copy of it can survive through another.

But even after having admitted kin selection, we still have a gaping hole in any attempt to explain altruism. If, for example, I help a blind man cross the street, it is plainly unlikely that I am being prompted to do this because he is a close relation and bears my genes. And the animal world is full of all sorts of elaborate forms of cooperation which extend far beyond the boundaries of mere relatedness.

Game theory—specifically one particular solution of Prisoner's Dilemma—suggests an answer: We cooperate because it is the most beneficial strategy. The most successful solution yet devised to all versions of Prisoner's Dilemma is known as "benign tit-for-tat." The first time we play, we expect the best of our partner. If he lets us down, we respond by noncooperation the second time; if he cooperates, we cooperate the second time. And so on. Running this through a computer millions of times suggests this is an "evolutionarily stable strategy"—it produces a benign environment in which everybody thrives. In other words: In the long run, simple savagery or hawkishness is not the most beneficial strategy and is not, therefore, likely to be the outcome of evolution. Some form of cautious altruism or dovishness, however, is.

Clearly one solution to one game is not likely to represent any kind of conclusive addition to evolutionary theory. But the point about this solution is that it shows the way the mathematical modeling involved in game theory can provide forms of explanation that are not intuitively obvious. Intuitively, for example, we might think the selfish gene would be bound to result in selfish organisms; Prisoner's Dilemma merely shows why this might not be so.

More importantly, the mathematics of game theory provides a logical bridge between evolution and behavior. Without such a bridge, explaining behavior through evolution is simply one of those backward narratives about adaptation that is always open to the criticism of circularity—this behavior is the way it is because it is the way it is. In that case, the fact of natural selection is the only bridge, and, as in the case of altruism, it has to be twisted into some fairly peculiar contortions to fit the facts. Game theory reveals that the contortions may not be so peculiar after all, but rather simply the most well-adapted outcome of millions of moves by millions of organisms, conscious or unconscious. It suggests a mathematical inevitability about types of behavior we call "moral."

But, most important of all, it turns the narrative of evolution into a moral narrative. Why are we nice/kind/helpful to each other? Because these have been found to be the best moves in the evolutionary game. (I am simplifying enormously here. Obviously we are often nasty toward each other, but the only points I wish to make are the simple ones that game theory joins evolution and behavior and that it explains morally diverse responses.)

That, in essence, is the basis of evolutionary psychology. It is a radical development because it embraces what philosophers call "the naturalistic fallacy," which claims that moral direction can come from nature—that "ought" can be derived from "is." Clearly evolutionary psychologists are not going to say we always should do whatever comes most naturally; the result would be mayhem and, in any case, the word "natural" is confusing. A man may feel it is "natural" to have sex with every woman he finds attractive and act on that basis. But, in fact, many evolutionary psychologists would say that monogamy is a "natural" condition because it best accords with our evolutionary interests. Having sex with every attractive woman might, therefore, be profoundly unnatural. Nature might not provide instant and obvious guidance but, in the form of evolutionary history, it is likely to provide moral and psychological understanding—an understanding that cannot be attained by any other method. Evolutionary psychology is also radical because it seems to offer precisely the secular, rational basis for morality that has eluded philosophers for three hundred years.

Fans of evolutionary psychology are in no doubt about the historic importance of this discipline. Robert Wright, whose book *The Moral Animal* has become the standard popular text of this new discipline, is convinced that we are witnessing a "paradigm shift," a fundamental readjustment of scientific perspective which will lead us to deep human truths. "For example," he writes, "can a Darwinian understanding of human nature help people to reach their goals in life? Indeed, can it help them choose their goals? Can it help distinguish between practical and impractical goals? More profoundly, can it help in deciding which goals are worthy? That is, does knowing how evolution has shaped our basic moral impulses help us decide which impulses we should consider legitimate?

"The answers, in my opinion, are: yes, yes, yes, yes, and, finally, yes."

Evolutionary psychologists, says Wright, are seeking out a deeper unity within the human species. They note the way guilt and the need for social approval are common themes in all the various human cultures. They further note that each individual's degree of guilt and need for social approval varies. The trick, as far as individual psychology is concerned, is to analyze these variations in terms of the reaction between the individual's genetically determined impulses and the environment—which, as I have said, though Wright doesn't, is comprised largely of other people genes.

Here we can see that evolutionary psychology is moving decisively into the territory once occupied by Freud. Wright is explicit about this, saying about evolutionary psychologists that "in contrast to Freud, they don't have such a simple, schematic view of the human mind. They believe the brain was jerry-built over the eons to accomplish a host of different tasks."

Freud's view was, in these terms, an insufficiently well-informed assault on human nature. Certainly he was right to place biology at the center. But his biology was essentially groundless; his psychodramas seemed more important than the underlying organic reality. He was also right to speak of a subconscious. One of the central insights of evolutionary biology is that the forces of evolution have found it necessary to conceal our natures from our conscious selves. It may sometimes, for example, be necessary to think we are being noble so that we may better serve the promptings of our genes. A high degree of self-delusion is likely to be built into the psychologically evolved brain. As a result we do have a subconscious but it is not defined by the systems of repressions of which Freud spoke; rather, it is simply the realm of naked genetic self-interest which often pursues its goals by deceiving the conscious mind about its true motivations.

But, remember, the original opposition to sociobiology was political. It was seen as a covert means of scientifically validating a strongly conservative view—a way of saying that things were the way they were because of nature and, therefore, should not, perhaps could not, be changed. It rooted human nature deep within us rather than in the social environment and, as a result, tended to remove the urgency from demands for social change.

In fact, in the writings of Matt Ridley, an English zoologist who emerged directly from the selfish gene culture of Richard Dawkins, this conservatism becomes explicit. In his book *The Origins of Virtue*, Ridley uses a gene-centered view of human nature as the basis of a political program designed to right the wrongs of liberalism. "If we are to recover social harmony and virtue," he writes, "if we are to build back into society the virtues that made it work for us, it is vital that we reduce the scope and power of the state."

Our evolved natures mean that we are predisposed to flourish in a free market. Ridley continues:

> For St. Augustine the source of social order lay in the teachings of Christ. For Hobbes it lay in the sovereign. For Rousseau it lay in solitude. For Lenin it lay in the party. They were all wrong. The roots of social order are in our heads, where we possess the instinctive capacities for creating not a perfectly harmonious and virtuous society, but a better one

than we have at present. We must build our institutions in such a way that we draw out those instincts. Pre-eminently this means the encouragement of exchange between equals. Just as trade between countries is the best recipe for friendship between them, so exchange between enfranchised and empowered individuals is the best recipe for cooperation.

The message of Wilson, Wright, and Ridley, as well as of many others, is that the gene-centered view of the world has given us the means of knowing something we have never known before: the truth about human nature. Human science may be on the verge of becoming more firmly based than ever before. The mighty gene offers a new way in which we can be said to know ourselves, a new way of realizing an ancient aspiration.

In the earliest days of the Western tradition, Socrates said, "The unexamined life is not worth living." The highest good of the human mind is to examine itself and to know itself. Science, in the form of evolutionary psychology, gives this aspiration a new form. Instead of the nuances and interminable analysis of philosophy, the scientific method offers the prospect of the "reduction" of human experience to comprehensible, provable formulae.

As G. K. Chesterton would have spotted at once, the problem is that this new way amounts to a massive inversion of our previous moral foundations. Kant's morality rested on the impulse to the good within each individual—a place he could reasonably suppose to be beyond the reach of science. Christ's rested on the absolute of the human soul and its relationship to God—a place beyond the reach of any external wisdom and interference. But the soul, the inwardness, of evolutionary psychology is our biological history. And it is neither unknowable nor unreachable. Indeed, its highest virtue may be its legibility—for, by reading of the text of our history, we may be able to rediscover the moral center that many thought would be lost forever with the decline of religion.

This would be the supreme triumph of the mighty gene, placing it at the very center of our culture and our beliefs. It would elevate genetics and evolution to a Theory of Everything, equal to or greater than anything the physicists might produce. And, as John Vandermeer has written, "A theory that explains everything is sort of like God."

So the new God becomes DNA and his altar is the great bronze sculpture which adorns the lobby of James Watson's Cold Spring Harbor Laboratory in New York—the twin, open-ended spirals of the double helix, cool, unadorned, rising from nothing and going nowhere.

chapter 6

Seeds of Destruction

I have a niece named Fiona Appleyard. She is now, miraculously, twenty-nine. This is miraculous because Fiona suffers from a very rare and virulent form of muscular dystrophy, a disease that almost invariably afflicts males. Fiona is one of the very few female cases in the world.

She is the most extraordinary person I have ever known. The wastage of her muscles means that she has never been able to walk or, indeed, to make more than the most minimal movements. She has been bed- and wheelchair-bound all her life. When she was eleven she underwent two devastating operations to prevent her body from collapsing on itself. The first winched her spine into a vertical posture and the second fused the vertebrae to keep her upright. On the night before the first operation she asked her doctor if she would die. "You might," he replied. She could not speak when she came out of the operations, but, by a prearranged signal, she blinked twice to indicate she was "all right."

The disease leaves the intellect untouched, and Fiona has a formidable intellect. She is stubborn, short-tempered, fond of strong cider and

Indian curries, and prone to paroxysms of laughter, which, alarmingly, paralyze her features into an open-mouthed, closed-eyed grimace. She speaks with a high-pitched Scottish accent, though since she has not lived in Scotland for years, I suspect her impatient, ironic twang is another aspect of her stubbornness. She is currently writing her auto-biography and has announced the fact in letters to all the national newspapers.

Fiona can play the British National Health Service like a maestro, once engineering her own transfer by ambulance to Papworth Hospital near Cambridge so that she could try out a new breathing aid that she thought might improve her sleeping. It worked. Fiona is always right about her own treatment.

She is skeptical, tough, and optimistic. She has always made plans for the future—one, when she was a teenager, was to be an air hostess. I dreamed of ways of blackmailing British Airways into accepting her. I have occasionally seen her afraid, usually when a mild cold threatens to kill her because of the fragility of her respiratory system, and I have once seen her despair—when a cold meant she had, yet again, to go into the hospital and endure days of having phlegm sucked out of her. Her weakness means that her coughs are ineffective, so fluid simply accumulates in her lungs.

That last was the one occasion when I felt I truly did something for Fiona. Her willpower had collapsed at the prospect of another stay in the hospital. For a moment she seemed to be choosing the alternative—death. I said, in the angry, demanding tones I knew she would understand, "Fiona, listen to me. You've done this before, you'll do it again." Calling on her unimaginable reserves of courage, she once again roused her vast spirit and once again survived.

Muscular dystrophy is a recessive single-gene disorder. As such, it appears a bewilderingly simple affair. The gene involved—one of the largest so far found—produces a protein that is necessary for muscular development. Fiona lacks that protein, so her muscles simply waste away. Why not correct the gene? Because millions of copies of the gene in millions of cells would have to be corrected to have any effect and because, so far, such precision interventions at the molecular level have proved impossible. Why not provide the missing protein? Because it does not work; the biochemistry is not yet understood. The muscular dystrophy gene was identified in 1983, but for Fiona and her cosufferers, nothing can yet be done.

A couple of years ago I led a debate at the World Economic Forum at Davos, Switzerland, on genetics. We discussed muscular dystrophy, and one woman, an Australian doctor, said she was proud to have prevented a number of cases. She had done so by detecting the condition prenatally and aborting the fetuses. Abortion is, for the moment, the only decisive response to the disease. I use the word "response" because I am not sure "treatment" would be quite right—the patient is, after all, eradicated, not cured. I have sympathy for that doctor. Contemplating the lives of sufferers—"They sit around in their wheelchairs waiting to die," she said—might well make one proud to have prevented more such lives. But then I think of Fiona and I am not so sure.

She might never have been born. Prenatal screening might have meant that her parents could have chosen to keep only a healthy child, Fiona's brother or sister. And why not? Every individual is just a chance emanation from the gene pool. Why not choose a healthy chance rather than a diseased one? I could scarcely know that a three-month-old fetus was destined to be stubborn, curry-eating Fiona. I could not reasonably argue, therefore, that, by stopping an abortion, I was saving Fiona. In addition, the emotional and, to both the family and society, the financial savings would have been enormous. Many lives would not have been distorted by the demands of the disease. And yet my life—and the lives of others who have known and loved her for what she uniquely is—would have been distorted without the experience of Fiona.

I have saved my discussion of disease and genetics and my story of Fiona until this late point in my book in acknowledgment of the fact that, for gene and science skeptics like myself, disease is the most serious challenge to our skepticism. Stop anybody in the street and ask them what they think is the most important contribution science has made to the world and they would almost certainly answer the conquest of disease. And I have delayed telling the story of Fiona because she says something essential about human experience.

The great French mathematician and philosopher Blaise Pascal wrote an essay called *A Prayer to Ask God for the Right Use of Sickness*. As the philosopher Timothy F. Murphy has written, this title sounds odd to modern ears because we think of disease and suffering as evils to be resisted, not as things to be put to proper use.

"There is, of course," writes Murphy, "absurd suffering and pointless difference. But it is also true that we do not know what our lives are worth if we do not countenance the price we would be willing to pay for them, the difference we would be willing to endure for them."

This goes to the very heart of what it is to be human. But first we have to acknowledge the simpler point that science has massively reduced the human cost of disease. The Black Death no longer threatens to eliminate a third of our populations as it did during the plague years in Europe. Our children are not being crippled and killed by polio. Smallpox has been eradicated. Tuberculosis, though now on the increase again, seldom ravages contemporary lungs. Public health measures arising from scientific knowledge have, by suppressing infection, almost doubled the average human life span in the developed world over a period of less than a century. Infant mortality rates have fallen spectacularly. We live longer and, thanks to modern analgesia, we suffer less pain in dying.

Science, through medicine, has provided precise quantifiable benefits for humanity. We may quibble about whether the personal computer or the Boeing 747 has made us happier or better people; we may, in an antiprogressive mood, claim we would be better off without them. But it would be infinitely more difficult to say we would be better off without the polio vaccine, or that a bracing dose of bubonic plague would strengthen our spiritual lives. Through medicine, science provides its most intimate, most exact benefits.

The crucial next step in medicine will be genetic. In some ways, medicine without genetics is coming up against the limits of its power. "In the past fifty years," Richard Lewontin points out, "only about four months have been added to the expected life span of a person who is already 60 years old." In other words, we cannot necessarily expect further improvements from traditional medicine.

Genetics offers a way out of this impasse. It promises to take medical science to a whole new level of competence. All medicine until now has been based on a high degree of ignorance. Our drugs are happy accidents—compounds found mainly in nature that just happen to work. Our surgery is the refined product of a long and bloody history of trial and error. In all areas of medicine we have been trapped on the surface of things, always uncertain about the complex biochemistry of which the organism is formed. But genetics seems to offer the possibility of getting to the root of that biochemistry. It holds before us the possibility of complete knowledge and, therefore, complete power over disease.

It may offer a new way to defeat the most feared killer of all, cancer. Years of effort and billions of dollars of research funds have achieved surprisingly little. In spite of many advances in the treatment of cancer, the disease remains intractable and the primary existing therapies—radiation and chemotherapy—can themselves cause appalling suffering. They are, as one geneticist has put it, "no way to treat a human being." All cancer is ultimately genetic in that it involves a change in the DNA. "We contain within us the seeds of our own destruction," Michael Bishop, of the University of California at San Francisco, has said. He added: "Cancer is within our genes." Genetics would seem, therefore, to offer the best, perhaps the only, hope for a new approach to the disease.

Defeating disease is the primary inspiration of most of the genetics being conducted in the world today. As Robert Cook-Deegan graphically puts it:

> A geneticist can work for years in a laboratory, never seeing an afflicted patient or commiserating with an afflicted family. The daily laboratory routine is relatively stable, if intense and demanding. Once the impact of a disease is directly experienced—the pain and devastation it causes for specific people—laboratory work acquires a new meaning. It demands greater urgency. The stakes go up; the room for excuses and tolerance of delay go dramatically down. Laboratory manipulations become less an exercise in abstract problem-solving and more a holy crusade against a common enemy. Disease becomes evil; eradicating it a primary need. Medical research

differs from other scientific fields in this respect. It is driven by this passion for life—the hunger to understand life in order to preserve it.

"I think the main reason for our work in the Human Genome Project is a medical one," said Francis Collins. "The reason is to alleviate human suffering. It's a natural extension of biomedical research. And, if you can't argue on that basis, then it's difficult to defend on the basis that it's good basic science or good for the economy. The real reason is to do something about large numbers of diseases that have remained obscure and difficult to diagnose and impossible to treat. This is not just for rare diseases but for virtually every human disease that has a genetic component and that means essentially everything except trauma."

The science of genetics is, overwhelmingly, the science of disease, primarily of those "inborn errors of metabolism" that cause human suffering. It is a continuation—the only logical one, say many geneticists—of that medical tradition that has done so much to overcome the shortcomings of our biological natures. Indeed, it has done so much that we expect much more. Medical miracles seem to be a kind of contemporary birthright; we expect a cure from our doctors in the same way we expect a mechanic to fix our car. Deep down, said one New York friend of mine, Americans think that death is optional. And they think that way because they have become so used to the idea that the body, like any other machine, can be fixed.

Immortality, of course, is, for the moment, asking too much. But perhaps it is not asking too much for death to strike swiftly and painlessly after a lifetime of good health, to make life as good as possible and death as easy. As Francis Collins said: "The ideal circumstances are where everybody is healthy until one day they suddenly and very definitively drop dead. Then you have no need for a health care system at all. You don't want people to have long, lingering illness; you want them to go quickly."

Genetic developments are, undoubtedly, aimed at arriving at this utopian state of affairs. But they have deeper implications than the mere prolongation of human life and the alleviation of individual suffering.

There is one problem that should be stated immediately—the problem of an excess of genetic enthusiasm. David Suzuki and Peter Knudt-son have written of "the incredible eagerness with which scientists and nonscientists alike have historically grasped at quick, socially convenient definitions of genetic disability or disease." And they add: "Like some prescientific shamanistic healer who finds people's fortunes revealed in fallen strands of hair or the discarded clippings of their fingernails, we unconsciously begin to extrapolate from the part the qualities of the entire organism."

The mighty gene is seen as medical salvation. Precisely because genetics is seen as the next big step in medicine, we start to believe it is the *only possible* step. Clearly such a belief opens the way to all kinds of possible error. Suzuki and Knudtson illustrate their point with the story of XYY men. Men are the most violent sex, and it has been assumed this tendency can be traced to the genes. In the sixties it was found that some men carried an extra Y chromosome, giving them an abnormal complement of one X and two Y chromosomes. Since it is the Y chromosome that determines maleness, people began to think that this additional dose of maleness may account for particularly violent behavior. For a decade this became, both in scientific and popular circles, established wisdom—there was even a television thriller series based on the idea. But, by the end of the seventies, it had become clear that there was no certain basis for the idea. A genetic myth had been constructed on the basis of flawed knowledge or complete ignorance.

With that in mind, we can turn to the first big theme of medical genetics: prediction. I was, rather chillingly, told by a geneticist who had just given me tea that a sample of my saliva from my cup could already be used to discover what I was likely to die of. This is, assuming I carry one of those conditions that can easily be genetically identified, true. Once we understand how genes work, we can, it is hoped, make specific forecasts about the medical futures of individuals. This could lead to a revolution in medicine. Currently we treat a disorder when it occurs. Preventive medicine through vaccination and health advice is mainly aimed at populations rather than individuals. But genetics may make it possible for individuals to take specific preventative action long before any symptoms appear.

This idea gets people very excited. Joseph Levine and David Suzuki quote the geneticist Leroy Hood in their book *The Secret of Life*: "My God, once we've identified the genes that control these traits and understand their function, we'll be able to repair their defects. And what we'll do is when you're born, we'll take your DNA and look at a hundred disease-predisposing genes, and we'll identify your pattern of predisposition for disease, and then institute appropriate regimes so these things will never come up. Once we're able to do that, we've revolutionized medicine by these preventative techniques. I see this as the real key to gaining control of escalating costs. That's the future and it's very exciting."

Hood's vision, like Collins's perfect-health/perfect-death scenario, is that genetic knowledge will actually remove the need for medicine as it is currently understood. Rather we shall simply be provided with an analysis of our genetic predispositions and told the correct course of action for the rest of our lives. This might mean avoiding—or, indeed, *not bothering* to avoid—certain activities like eating animal fats, smoking, or exposing yourself to certain forms of pollution. Or it might mean drug treatments, or even, if it works by then, gene therapy to counteract the effect of bad genes. In the case of infectious disease, it might mean exact genetic adjustments to our immune systems.

It should be made clear at this point that we all have bad genes. The average, healthy person is statistically certain to carry a number of recessive alleles—estimates vary between four and twenty—that would, if paired with similar alleles, result in catastrophic disease. If we spread the definition of bad genes a little wider to include predispositions to illness, then all our genomes will be found to be massively flawed. In the case of recessive single-gene disorders, the possession of one allele will have no effect on the individual, though it could have significance when it comes to deciding with whom one chooses to mate. In the case of predispositions, this may have a direct effect on the individual, but such an effect will be expressed as a probability. For example: You have a 43 percent chance of having a heart attack by the age of forty-five; therefore you should exercise, avoid animal fats and tobacco, and, perhaps, take a blood-thinning aspirin a day or certain drugs to reduce your cholesterol levels. Knowing the specific susceptibilities of your body allows you to plan your life to give you the best possible chance of achieving the perfect-health/perfect-death outcome.

But, of course, putting it this way immediately reveals a certain flaw in the utopian vision: People don't listen. Consider smoking.

"The biggest thing you could do to improve people's health," said Peter Goodfellow, "would be to stop them from smoking—people know this and yet they go on smoking."

These days, smokers are harangued, harassed, and abused. They must, by now, know the risks involved and that they are subject to a high degree of social rejection. And yet many people still smoke. They refuse to act upon the clearest possible predictive information. Many also become obese through eating too much in spite of equally clear predictive information. In the case of smoking, one can say that addiction is involved. But that is not really the case with fatty foods; they do not, as far as we know, create chemical dependency. Fatty foods and tobacco are demonstrably bad for most people, yet they are still consumed in enormous quantities. Clearly, telling people how to manage their lives doesn't always work. As a result, there is a significant gulf between the medical utopia foreseen by geneticists and the way real people behave.

This then raises the question of blame, a correlative of the question of guilt I discussed in the last chapter. If I smoke in spite of all the health warnings, should I be held responsible when I am struck down by a smoking-related disease? Currently the answer is, in one form or another, yes. Smokers are discriminated against through insurance policies. With a state health insurance system like Britain's, in which treatment is free at the point of use, such discrimination does not, in theory, happen. In practice, however, there have been a number of cases where doctors have said heavy smokers are less deserving than nonsmokers. Such attitudes have not yet been raised to the level of policy. But, as pressure on state health spending increases, such discrimination is likely to become more frequent.

But what about discrimination against somebody who refuses to lose weight? Again the logic might be that they are to blame and their free choice should not be a financial burden to others—either taxpayers or other insurance policy holders who might see their premiums rise as a result of their companies having to provide coverage for people who are irresponsible about their diets. Okay, we might say, you are not morally wrong to be fat, but it would be immoral to ignore the fact that your conscious decisions are putting you at risk and, therefore, you cannot expect others to bear the cost of your health care.

Genetic forecasting adds a new dimension to this problem. Clearly we cannot blame somebody for having a predisposition to heart disease in the same way we might blame them for smoking or overeating. But, equally clearly, they still represent a burden to others who have no such predisposition. They ought to be, in strict financial logic, discriminated against. This represents one conception of fairness—fairness to those without the predisposition. Yet, since those with the predisposition are innocent of any contribution to their condition, it might be just as fair to say they should not be discriminated against. This would seem to be a higher kind of fairness in that it treats everybody in the same way, whatever their genotype.

But the idealism of such a position is not easily translated into the real world. Insurance companies often decline to cover "preexisting conditions." Is a genetic predisposition a preexisting condition that can, under the same rules, be excluded from coverage? And some companies have tried to deny insurance coverage to children born with genetic defects after their parents refused to abort when they were warned by prenatal testing. The Clinton administration has just banned discrimination based on genetic testing. But the underlying pressure remains.

The insurance industry is tying itself in knots over this. One insurance spokesman has said he wished genetic technology had never been invented, so complex are the problems it creates. The solution of the geneticists I have spoken to in the U.S. is twofold: There should be a legal right to DNA privacy (nobody can discriminate against you on the basis of your genome, just as now nobody can discriminate against you on the grounds of race) and a transformation of the American health insurance system into one more closely resembling Britain's.

"We're heading," said Francis Collins, "for a real train wreck on the health insurance issue in the U.S."

The danger, if this train wreck is not avoided, is that genetic information will prove as divisive in health care as it may in politics and sociology. An uninsured and uninsurable "genetic underclass" may be created consisting of those whose genomes are considered just too dangerous. Or, as I shall explain, the cost of genetic testing could lead to the medical reinforcement of economic divisions. Lori B. Andrews, a lawyer, writes:

> The expense of genetic tests prevents many people from using them. Prenatal diagnosis, for example, is mainly used by women from the middle and upper classes. This disparity has implications for public policy far beyond the area of genetic testing per se. Prior to the advent of prenatal diagnosis, a child with a genetically based mental or physical disability could be born into a family of any socioeconomic status. Middle-class and wealthier families used resources and connections to lobby state legislatures to pass laws providing for adequate education for children with disabilities. With the use of prenatal diagnosis and abortion, fewer such children are being born to couples of higher socioeconomic status. Affected children may become—like "crack" babies and border babies—an issue for the poor, with many fewer protections and resources available to them.

At heart the issue is the definition of disease. Genetics can tell us we are, in some new sense, ill before we feel any symptoms. It can tell some symptom-free people that they are very ill indeed.

"The result will be," the philosopher Dan W. Brock has written, "that people who feel healthy and who as yet suffer no functional impairment will increasingly be labelled as unhealthy or diseased. . . . For many people, this labelling will undermine their sense of themselves as healthy, well-functioning individuals and will have serious adverse effects both on their conceptions of themselves and on the quality of their lives."

Clearly, if I have cancer, I am sick. But am I sick if I have a predisposition to cancer? And, as I say, we all have bad genes, so are we all sick? Is sickness, in some sense, our natural condition—a form of orginal sin—and is health a distant ideal to which we can aspire but which we can never hope to achieve? If that is the case, then genetics, far from reducing health care costs, could increase them massively by setting impossibly distant goals of perfect health. We could then spend ever increasing amounts of money on an inevitably doomed attempt to attain an ideal of physical perfection.

To some extent our culture has already prepared us for this. Look at the plastic surgery industry, which is based on the idea that we should constantly strive for some physical ideal and try to thwart our natural predispositions. And consider the form of the whole popular fascination with health. There always seems to be more we can do to improve ourselves. Health has already ceased to be negatively defined as the condition of not being sick and is now regarded as an ideal human condition, the attainment and preservation of which requires positive effort. In this sense popular culture has anticipated the insight of genetics that we are all inwardly sick.

But the immediate practical problem is that we will not move smoothly from a condition of genetic ignorance to a condition of complete knowledge. Rather, there will be a disconnected stream of information coming from the laboratories as more and more of the human genome is sequenced. As a result, some conditions—muscular dystrophy and, perhaps later, heart disease—will be definitively identified as genetic, while others, say, schizophrenia and alcoholism, will be assumed to be genetic until researchers can come up with some clear indicators. This will lead to distortions in the health and insurance markets. Consider the sheer oddity of the fact that the alcohol industry is keen to find the genes for alcoholism; if they do, it frees them of blame and may sell more drinks to those without the disease.

The further and perhaps more painful problem is that effective treatment will continue to lag far behind genetic information. Isolating the muscular dystrophy gene is, for the moment, pure knowledge; it has no practical implications for those born with the disease. But, of course, it does have one big practical implication for those not yet born: They can be aborted.

Abortion is currently the one sure "treatment" that has arisen from genetics. Whatever one's feelings about abortion, everybody must agree that this does not amount to a very positive achievement for this new form of medicine. Jerome Lejeune was the scientist who discovered that Down's syndrome was caused by the condition known as trisomy 21, in which some children are born with three, rather than two, copies of the twenty-first chromosome. His motive was to find a cure, but, to his great distress, his discovery led only to thousands of abortions. The strange situation has thus arisen that, in cases where abortion is the only available option following a pre-natal diagnosis, medicine has put itself in the position not of treating a patient but of ridding a family and the society of a particular burden.

This is how the philosopher Philip Kitcher puts the problem: "When the only option is to terminate a pregnancy, there is no such identifiable person whose lot becomes better. Possibly, by intervening, we would make the world happier than it would otherwise have been. . . . But we have not brought any benefit or relief to the bearer of the foreshortened life." In short, abortion is a unique medical treatment in that it helps everybody except the patient.

The option of abortion also raises the difficult question of the seriousness of any given condition. At the moment, the objects of pre-natal screening are largely serious conditions. This is, in part, because these are the diseases we know how to detect. But it is also because the two primary methods of screening—amniocentesis and chorionic villus sampling—are relatively radical and slightly risky procedures. Not everybody, therefore, is routinely screened; only those with a known hereditary risk are. But screening is almost certain to become easier and risk-free. Intense research is now being directed at the possibility of taking a sample of the mother's blood and using any fetal cells that happen to be circulating for DNA screening. This method would mean that virtually every pregnant woman would be routinely screened. Prospective parents will, as our knowledge of the human genome increases, be deluged with genetic information about their children. How will they act on this information? How will they define disease or an unacceptable genome?

Few would argue that muscular dystrophy, Huntington's disease, or cystic fibrosis are very serious illnesses, and many people will find perfectly understandable the decision to abort an affected fetus. But what about a predisposition to heart disease that may only affect the child in his thirties or forties? What about a tendency to aggression? What about homosexuality? What about brown eyes when the parents want blue? Clearly a scale is implied by such questions—a scale that runs from definitely serious to probably trivial. But that then raises the issue of who decides what is trivial. Most people would regard brown eyes as a trivial reason for abortion. But how many would say homosexuality was trivial? This, in turn, introduces a complex moral relativism into genetic decisions. There is no sanctity attached to the individual; rather, he or she becomes a collection of characteristics, each of which can be judged on some scale of relative significance. At this point it becomes difficult to distinguish human beings from consumer goods.

Geneticists and doctors are concerned about this, though they usually fall back on the freedom of the individual to choose. They insist that their advice must be "nondirective," which means they give their patients the facts and leave them to make the decision.

Francis Collins, a devout Christian, says: "I have some personal difficulties as a physician who counsels parents about DNA testing for some possible affliction. There are times when parents choose to terminate a pregnancy for rather modest reasons. I feel uneasy about having participated in that, but I believe that the genetic counselor must be nondirective in such situations, and the parents would never know I felt uneasy."

But he is reluctant to use his "feeling uneasy" as the basis of any kind of dogmatic insistence about right and wrong. "I'm aware of the incredible agonies that people go through," Collins says, "when confronted with this information, trying to decide whether to take their chances or whether this is a blessing from God in a very unusual form. I have trouble with people who come down dogmatically on either side of this."

The question becomes: Is there a higher court of appeal than the individual conscience? This is clearly a question that can be applied to every aspect of social and political life, and it tends to produce different answers depending on what sort of issue is at stake. If a country goes to war, it may allow conscientious objectors to avoid conscription, but the overwhelming assumption is that, in this extreme situation, the interests of the collective are more important than the qualms of the individual. Equally, society does not allow an individual to act upon his conscience if it directs him to do something plainly antisocial—say, to kill a man who burgled his house. The consciences of many devout Muslims may direct them to assassinate the novelist Salman Rushdie because such an action was legitimized by the Ayatollah Khomeini's death sentence for Rushdie's supposed blasphemy in his novel *The Satanic Verses*. But non-Muslim societies cannot allow freedom to such an expression of conscience.

In general, however, advanced secular democracies find it difficult, if not impossible, to sustain belief in any higher moral law than that of the individual. There is no religious or moral accord; there is only a vast and frequently contradictory plurality. For many this is unarguably good news. It celebrates freedom and places our own thought processes at the moral center of the world. And it is certainly the belief that forms the basis of almost all the current thinking on how we deal with genetic information.

Counseling is at the heart of this process. At first, between the 1940s and 1960s, this was largely done by physicians and geneticists. But, in the seventies, a new type of expertise was established. Genetics, because of its complex, intimate predictive nature, produced a whole generation of people trained as counselors—explainers of the science in terms the lay person can understand. The basis of what they do was laid down by Carl Rogers—"the father of 'client-centered counseling' "—who defined the role of the counselor "as clarifying and objectifying the client's own feelings." In other words, the counselor was not there to impose a specific view or decision, but, rather, to place the information before the client and then to lead that client to a decision which embodied his or her own beliefs. This is the essence of "nondirective counseling." It is based on the belief that there is indeed no higher court of appeal than the individual conscience.

Philip Kitcher describes a utopian ideal of what nondirective counseling should be: "Counselors will sometimes be the last and only possibility for articulating a course of action for people, whose awareness of their own predicaments and options is, at best, partial. They will have to learn how and when to elicit the decision that most accords with their client's interests and values, neither inserting their own priorities nor remaining so remote that the advice they give is useless."

Of course, it is easy enough to see the flaws in the idea of counseling as nondirective. First, it is a hard position to maintain, as the inclinations of the counselor will always tend to surface. Second, the idea of "clarifying and objectifying" the feelings of a client when he or she has just been confronted with an utterly new form of knowledge is questionable. Most likely clients will not come up with any form of internal moral response and will seek desperately for hints about what to do from the counselor and take their cue from whatever hints they might pick up or imagine. Third, the very fact that they are being counseled indicates that there is a problem; a screen has been established to catch a particular genetic condition.

"In social terms," as Troy Duster explains it, "the mere existence of a screen assures that the 'idea of a screen' will penetrate the consciousness of lay persons, and communicate that the screened-for characteristic is undesirable." The client is forced to reason: I wouldn't be here unless something was wrong. The very fact that it is seen as wrong compromises the ability of the counseling to be entirely nondirective.

The pressure to have tests and to act on their results can be enormous. "Genetic testing," Dorothy Nelkin has pointed out, "is encouraged by legal pressures, such as wrongful-birth and wrongful-life suits against physicians who neglected to offer their pregnant patients tests that could predict fetal disorders. If tests are available, they will be used." Also, it is hugely persuasive for counselors to tell people that they *ought* to know to fit in with the medical procedures. Diana Fritz Cates, an ethicist, has written, "A counselor should help a person who is at risk to come to see that, if she waits to be tested—if she waits until known symptoms begin to arise—it may be too late at that point for her to experience a coming to terms with herself."

But that phrase, "coming to terms with herself," can be highly ambiguous. Here are the words of one young woman at risk from Huntington's disease: "There is no treatment or cure for HD. What good would it do me to know now? There was nothing I could do to change the inevitable one way or another. Would I really modify my behavior or lead my life any differently? A yes answer to that question would surely nullify the meaning of my present life. I decided that living my life to the fullest, with hope for the future, was the best possible solution."

The point is impeccable. What does "to come to terms with" mean? Does it mean changing your life? If it does, that must mean there is something wrong with your existing life. You already know that one day you will die of something. A balanced life must surely be lived with that certainty in view. If being told you are going to die of one particular disease at one particular time leads you to change your life—other than in the most trivial, practical details—then, clearly, you must be leading an unbalanced life. A basic change in your life in response to this type of bad news implies a criticism of your spiritual status.

Like the counseling profession, the current state of most abortion laws is based on the view that there is no higher court than the individual conscience. In Britain, the overwhelming number of abortions are for "social"—nonmedical—reasons. Asian families that do not wish to have girls can abort female fetuses. If they are allowed to do that, can there be any consistent argument against aborting a fetus that has brown eyes?

Those in favor of totally free access to abortion will not find these questions difficult. The rights of the individual will come first in all circumstances. But I think this position becomes inadequate when we take into account the wider social consequences.

There can be no doubt, as I have said, that the widespread elimination of fetuses regarded as undesirable amounts to a new form of eugenics. This is not the old eugenics of state intervention to improve the quality of the nation's gene pool. Rather, it is a form of privatized eugenics in which millions of individual decisions will change society.

And that is the key problem with privatized eugenics—it amounts to a judgment on the existing human population. And that judgment is profoundly conditioned by the current attitudes of society. People may abort homosexual fetuses because they don't like the way homosexuals behave and are treated in the world. But, instead of aborting them, we could simply improve the world, make it more tolerant of homosexuals. And what about Down's syndrome? Those afflicted tend to be loving and rewarding children, but they will be mentally retarded and they look different. They are, therefore, a social problem both because of the help they need and because they will be perceived as diseased. This becomes an even more painful issue when it is generally believed that women do not have to have Down's children. They can be tested and they can choose to abort, so having such a child can be seen as a decision which can be condemned. This is not even fair in its own terms, as most Down's babies are born to mothers not known to be at risk, so they are unlikely to have been offered a test. But the possibility of prejudice remains. As Professor Kay Davies put it: "Genetics will alter people's view of handicap. People that do bring handicapped children into the world will be looked upon as foolish and irresponsible. The eugenic impulse is very strong. Already people look at children with Down's syndrome and they can't understand why the parent didn't have the test. But quite a lot of Down's syndrome children are born to women under 25—so there is no point in them having a test."

Simpler, universal testing will, however, neutralize that last point. People will routinely be told, or at least they will have the opportunity to be told. Every Down's baby at that point will become a clear decision. And the decision to have the baby will be seen as a cost to society. For example, it costs around $60,000 a year to keep someone with cystic fibrosis alive.

The genetics of disease thus creates a pressure for normality, and this pressure is intrinsically discriminatory because it amounts to a negative judgment of the abnormal people we see around us.

As Evelyn Fox Keller wrote: "There are many problems associated with the geneticization of health and disease, but perhaps one of the most insidious is to be found in its invitation to biologically and socially unrealistic standards of normality, threatening not a return to the old eugenics, but a new eugenics—what the Office of Technology Assessment itself has called a 'eugenics of normalcy.' "

Elisabeth A. Lloyd, a philosopher, makes the further point that if we are to embrace normality as a medical program, we should try to find out what it is. "In fact," she writes, "molecular techniques should be understood as offering an unprecedented amount of social power to label persons as diseased. Hence, it is more important than ever to gain insight into the normative components of judgments about health."

Timothy F. Murphy compares the predictive power of genetics to the ancient art of haruspicy, or divining the future through the examination of animal entrails. "Certainly, we hope at least to be able to foretell the genetic future of particular individuals," he has written. "Such a hope raises important questions regarding identity and difference. To what extent will the genome project generate new classes of human inferiority? Will the genome project generate a theoretical subjugation of genetically atypical people, born and unborn, and thereby establish difference as disease or disability? Will the genome project mark difference as an undesirable trait and justify its eradication?" And he adds: "The moral significance of the project may prove, therefore, to lie in its significance for the interpretation of health and disease, normalcy and difference."

There are two steps involved in this process. The first is to characterize disease as genetic, the second is to characterize as a disease every abnormality that is genetic. Plainly, if we follow that logic, the outcome will be, via the processes of privatized or "domestic" eugenics, a population that is normalized according to whatever is perceived as normal at the time the technology becomes available.

This already happens with some children who are growing up unusually short—not so short that they would be said to be suffering from dwarfism, but short enough for their parents to worry that they might be at a disadvantage. Fearful for their children's future, they have them treated with growth hormone, now widely available thanks to genetic engineering. Is being short an illness or is it simply a human attribute that happens to attract discrimination? And, in any case, if growth hormone eliminates one level of shortness, then those at the next level will simply be regarded as unacceptably short. David Botstein made the point to me that an abnormality at the molecular level had, in fact, been found in many short people, so, arguably, it was a disease in the traditional sense. But this is stretching a point and rests upon the dubious supposition that every genetic abnormality is a disease. The fact remains that abnormality is defined by society, not by the genome.

As Camille Limoges puts it: "In a Darwinian world there is no longer any norm for a species beyond its relations to its environment, physical or biotic. Normality has no stable referential meaning: *the reference is in the experience* of the individuals in the population. The notion of 'genetic error' in that regard is a regression to typological thinking. . . . There is no such thing as a standard genome."

In other words, Darwinism and genetics destroy the ultimate scientific reality of the norm, but then, perversely, they give us the power to impose our own invented norms. This is yet another of those strange paradoxes, these curious inversions created for us by our new knowledge.

Many who are outside those invented norms would protest at this particular inversion. Indeed, activist handicapped groups have pointed out that the widespread abortion of handicapped fetuses amounts to a statement that the world would be a better place if they did not exist, a version of my own thoughts about Fiona.

Deborah Kaplan of the World Institute on Disability has spoken of prenatal testing as "a statement that disabled people shouldn't exist." In Britain, Tom Shakespeare, a four-foot five-inch sociology lecturer, made the same point when he said that aborting abnormal fetuses amounted to a statement that "people like me shouldn't exist." The Committee on Legal Affairs of the European Parliament has warned about the idea of regarding handicapped children "only as an avoidable technical error," a view that "undermines our ability to accept the disabled." On the other hand, can such an argument be used as a basis for limiting the freedom of parents to choose? Daniel J. Kevles and Leroy Hood do not think so: "It would seem to make little sense," they write, "to seek to preserve the dignity of one group by limiting the reproductive freedom of another. What would make a good deal more sense is to recognize that values of social decency compel us to live in a state of conflicted practices—endorsing the use of genetic information in personal reproductive choices while upholding the dignity of the diseased and disabled."

And David Botstein says: "Are we going to treat these people decently or are we going to treat them as pariahs? That is a question that is completely independent of the origin of the disease."

We should not, in short, allow knowledge to impose on decency.

Nobody could argue with that, of course, but it is an ideal that has seldom been attained in human history. Differences between people have often led to savagery, and genetics identifies very exact differences. Timothy F. Murphy has wondered whether new genetic knowledge will "permit the erosion of difference in favor of genetic uniformity, whether its characterizations will offer yet another standard of 'normalcy' to be used as the justification for the extermination of difference." In other words, by making ourselves more uniform, we become even more sensitive to difference.

As Kay Davies says, the eugenic impulse is strong, both for ourselves and for others. We want "good" children and, if we are prepared to make that effort, we tend to expect others to do the same rather than burden society with "bad" children. It is not difficult to imagine the kind of thought processes that might become commonplace. Just as cigarette smokers are currently seen as legitimate targets of discrimination because of what they do to themselves and what their smoke does to others, so those who give birth to handicapped children that could have been aborted may come to be seen as blameworthy.

But, of course, there are other, less philosophically fraught applications of genetics to disease. As I have said, a deeper understanding of the body's biochemistry should lead to more precisely designed and targeted drugs. Gene therapy should one day be available to correct the DNA of the afflicted. Little needs to be said about such methods. They may reasonably be seen as no more than extensions of the current medical procedures. As David Botstein says of gene therapy: "I'm just feeding the cells; what's the difference between that and feeding the gut?"

So our picture of medical genetics divides neatly into two: It can be seen as an uncontentious extrapolation of existing medicine—one further step in our glorious progress to the ultimate conquest of disease that raises no psychological, ethical, or social problems that are distinctively different from what has gone before—or it can be seen as a form of medicine that is quite different from anything that has gone before. It raises entirely new issues for insurers, counselors, doctors, and patients that cannot easily be fit into existing professional and personal categories. It threatens serious social divisiveness. And it transforms our sense of the identity and nature of the individual in a way that challenges our essential understanding of the meaning and purpose of our lives.

Early in this chapter I mentioned Pascal's essay *A Prayer to Ask God for the Right Use of Sickness*. The assumption of that title is that sickness has a function in our lives, that suffering might have a "right use." Today we find that almost impossible to believe. Our modern ability to overcome so many types of suffering has led us to construct a morality of health in which the eradication of suffering caused by disease is seen as one of the few virtuous projects on which we can all agree. This project has become, effectively, limitless, partly because we aspire to immortality and there is no logical limit to the number of years we might be able to add to our lives, and partly because our competence inspires us constantly to widen the definition of disease so that a whole range of human misfortunes or disadvantages can be "medicalized." Indeed, the current belief that everything, whether it is a disease or not, has a large genetic component encourages the view that it is fixable by medical means. And, if we don't fix it, there must be some reason why not—some reason that may well lead to us being blamed for our condition.

But what could Pascal have meant? Well, clearly he was writing at a time when there were few cures available. Little could be done about sickness, so it was reasonable to make the best of it; in the case of Pascal, "the best" meant the writing of some of the world's finest theological and philosophical literature. In addition, the Christian culture in which he was writing was one built upon a view of suffering as redemptive: Christ suffered and died for our sins. There was a transcendent connection between the trials of life and our fallen condition. This is expressed most literally in Christian Science, a faith in which disease is defined as an "error" and medical treatments are rejected in favor of a contemplative program aimed at eliminating the error.

Both the basic view of suffering as redemptive and the Christian Science view that it is a direct result of specific error are hard positions to maintain in the contemporary world. Yet they clearly mean something. They refer to a sense of depth in our experience, a sense of the difficulty involved in being fully human. They see being human as a hopelessly mixed experience, one in which happiness and ecstasy coexist with misery and despair, and in which health and disease seem to be inseparable aspects of our perpetually ambiguous place in the world.

And, hard as it may be to accept the tough prescriptions of such a position, it is equally hard to deny its truth. Even in perfect health, few people's lives are unambiguously good. Things go wrong, people die, we are plagued by doubts about our actions and place in the world, and so on. Whether or not you believe we are "fallen," it is clear we are not angels or gods, and equally clear that we never shall be. The Christian tradition teaches us to learn wisdom and compassion from this elementary fact of human life.

The medical tradition can do this as well. Wise doctors know there are few things they can really cure. My own doctor once told me that 90 percent of the people who come into his office have nothing wrong with them—nothing, at least, that medicine can treat. But the world is full of unwise doctors, whether in clinics or on confessional chat shows. They convince us that something can be done, should be done. They fill us with the conviction that all our ills can be cured. Genetics makes this worse by its apparent capacity to medicalize everything. It claims it will find the source of all our woes in our genes and, in doing so, persuades us that our woes are, indeed, illnesses. But we all have woes all the time. Are we, therefore, always ill? Genetics may, one day, cure some serious diseases, but meanwhile, it makes us sick all the time.

My niece, Fiona, has spent her life in a condition of sickness which few of us ever experience or could ever imagine. In knowing her, I have glimpsed what Pascal meant. Yet I know that long before she was, strictly speaking, Fiona—when she was a fetus—the science of genetics could have detected her disease and she could have been aborted. I now want Fiona to live. Would I then have wanted her to die? Or is that an illegitimate use of the word "her"?

The question then becomes: When she was a fetus, what was she? If, before there is a self, there is no person, then it would not have been Fiona that was lost in an abortion. Antiabortionists would deny this—most commonly because of a belief in the God-given soul that is already present in the fetus. But you could also say that, in being a potential human, the fetus is a part of the realm of human concern. The most avid prochoicer would have to admit that even if a fetus is not a person, it is not nothing; it is not completely without moral significance. My question is: What is this moral significance and can we do without it?

At this point the medical issue becomes the issue of the status and nature of the human self. So, at this point, I must try and say what I mean in the hope that I know.

chapter 7

The Spider
The Spider

> Tolstoy, Shakespeare, Dostoevsky, Kafka, Nietzsche have penetrated
> more deeply than John Buchan or H. G. Wells, or Bertrand Russell;
> but what we know on this level of half-articulate habits, unexamined
> assumptions and ways of thought, semi-instinctive reactions, models
> of life so deeply embedded as not to be felt consciously at all—what
> we know of this is so little, and likely, because we do not have the time,
> the subtlety and the penetration, to remain so negligible, that to claim
> to be able to construct generalisations where at best we can only
> indulge the art of exquisite portrait-painting, to claim the possibility
> of some infallible scientific key where each unique entity demands a
> lifetime of minute, devoted observation, sympathy, insight, is one of
> the most grotesque claims ever made by human beings.
> —ISAIAH BERLIN, *The Sense of Reality*

Edward O. Wilson writes of logotaxis, or "an automatic orientation
towards information." Like flowers turning toward the sun, the human
species turns toward information. Nothing more exactly characterizes
us than our urgent need to know—where the next meal is coming from,
the source of that strange noise, what the weather will do today, and if
this approaching stranger wishes to kill me. We are subject to number-
less threats, opportunities, and necessities, and we must subdue them
with words, evidence, information, and ways of containing and classify-
ing the chaos of experience.

Evolutionary theory can, I am sure, explain the phenomenon of
logotaxis. Long ago in the Great Rift Valley of Africa it probably
gave our ancestors the crucial competitive edge and, as a result, became
embedded in that fraction of DNA that distinguishes us from the apes.
It would have driven the formation of language as a means of sharing and thereby amplifying the effectiveness of information. It would
have sent self-conscious, thoughtful, calculating Cro-Magnon man out
of Africa into Europe and the Middle East to marginalize and ultimately eradicate the Neanderthals. This upright, sapient ape was
the first creature with the power to mold and form nature to its own

purposes. And it did so through the gathering and processing of information.

That is the sort of story we tell about ourselves these days. The terms "processing," "information," and "competitive edge" are in accord with the computerized, free-market temper of the times. It may well be a true story. Logotaxis would be explained simply as an exceptionally successful survival strategy. It has, undoubtedly, worked well. We have conquered the planet with information. So, apparently, the future is clear: We must know more; we must know everything. We must become what would once have been called gods.

But another, older story is also told about our logotaxis—that in the Garden of Eden, the Devil, disguised as a serpent, persuaded Eve to eat the fruit of the tree of the knowledge of good and evil and she, in turn, persuaded Adam. They ate, they knew—and God, in his anger at their disobedience, ejected them from Paradise. Now our thirst for information becomes Satanic; it distances us from God and condemns us to wander the earth, exiles from Eden, seeking salvation from our disastrously logotaxic selves. It is better, in this version, not to know what we are not supposed to know, to live in the "cloud of unknowing" that lies between us and God.

The evolutionary story explains our success; the Biblical story explains our failure. But they are both about the same thing—our compulsion to know. There is serious disagreement here. Version one says our curiosity makes us triumphantly human; version two says it is our downfall, that it deprived us of heaven on Earth. Which is right?

This is not a question that can be answered by assessing the likely truth of the two stories. There is obviously more evidence for the evolutionary version, and many if not most Christians will accept that the story of the Garden of Eden is probably a poetic rather than a literal rendering of the truth. But it is human, not literal, truth that is at stake here. So perhaps "Which is right?" is not even the appropriate question because the answer it demands is far too crude. The most ardent believer in the story of the Fall must accept that not all knowledge of the world is an offense to God; the most devoted evolutionist must accept that we are not improved by everything we discover. The answer to the question "Which is right?" is not one or the other, but something of both.

That is a consoling conclusion, but it is not really good enough. There is too much of a contradiction for Eden and evolution to live comfortably together in one mind. Nobody can really spend his life hovering contentedly between these two poles. And nobody does; as he veers slightly toward one, at once the other fades. Everybody I know is either Edenic or evolutionist in his outlook, not necessarily because he believes one story or the other, but because his temperament inclines him either to mistrust or to trust the fruits of the human will to know.

So the terms of the question and its absolute demands are right. It defines a human reality, for it is a question about purposes. "Which is right?" means "What are we here for?" Are we here to follow our lust for information wherever it leads? Or is there some other purpose, some justification or explanation of human life that might make us resist or contain our desire to know everything?

For most of human history such questions would have seemed meaningless. There was plainly no way we could know everything; how a simple flower grew was an impenetrable mystery. Yet, at the same time, total knowledge always seemed to be possible. There was no discernible limit to our ability to acquire information and to reason our way to explanations. The human mind could imagine anything. But the crucial, basic facts of the world were beyond our reach; the flower remained a mystery. There was always, therefore, a paradox—a gap between our knowing and our inability to use that knowing to explain. We seemed to be trapped in some intermediate state between godlike omniscience and blank ignorance.

"For in fact what is man in nature?" asked Blaise Pascal in 1660. "A Nothing in comparison with the Infinite, an All in comparison with the Nothing, a mean between nothing and everything."

Religion was the almost universal balm applied to the wounds inflicted by this inconclusive state. Mankind, according to one estimate, has produced 100,000 religions. Each has attempted to explain the peculiarity of our existence on this intermediate plane between gods and animals, between the infinite and oblivion. Each taught that our information mania was not enough. It could not reach to the core of existence. Certainly we were special, but not quite special enough. We were fallen, deluded, or deceived, tied to our sinful natures or to *sam-*

sara, the wheel of desire. Ultimate knowledge—the apprehension of God, the escape from desire—could not be had in this world.

And then along came science, a new way of resolving the paradoxical indeterminacy of the human condition. Suddenly our powers did not seem limited. Our logotaxis had turned us toward a new way of ordering information that promised to bring everything within the realm of reason. The mysteries implicit in any religious view could now be dismissed as merely provisional, only mysterious to the extent that they had not yet been invaded by the forces of science. Over time the realm of mystery would progressively shrink, eventually to nothing. To the hard scientific gaze, there could be no ultimate mystery, only the so-far-unexplored. The world and ourselves were logically completely knowable.

"I have no need of that hypothesis," said the French mathematician Pierre Simon de Laplace, with historic urbanity, when the emperor Napoleon asked him why there was no mention of God in his book, *Philosophical Essays on Probabilities*. The "need" for God had gone, but there was a price to be paid for this suave power that had made Him unnecessary. Religions are about human meaning, but science, this potential totality of knowledge, is not. The hard scientific view either leaves a world without meaning or a world with only one possible meaning— the meaning provided by the pursuit of scientific truth.

For some the very idea of meaning becomes, in the scientific age, an absurdity. Look at "the facts." Over the course of millions of years, unself-conscious life covers this planet, driven only by the need to feed and reproduce. Then the system driving this process—natural selection— arrives at the strategy of self-consciousness as a survival mechanism. Like many other products of evolution, such as the peacock's tail, this strategy gets carried away with itself and produces a degree of self-consciousness that seems, to its possessors, infinite. So, with comic pretension, this self-conscious being is stricken with wonder and self-questioning. He combines his feeding and copulating with love and belief and with mad, futile speculations about meaning and purpose. But these are not really as significant as they seem; they are just evolutionary contingencies, side effects of self-consciousness, or "epiphenomena," as biologists delight in calling what ordinary people think are the most important things in life. However "meaningful" they may seem to the thinker, they are, in reality, no more significant than the variant patterns to be found on peacocks' tails. Growing up, becoming a full subscriber to the scientific world view, means abandoning these speculations as obviously childish. There is no meaning; there is only knowing.

Depending on your way of thinking, this view can seem either incredible or the most obvious common sense. To the Stanford geneticist Dale Kaiser, the powers of the human mind are just too extraordinary to be explained away as a kind of evolutionary overspill. "It seems to me that much more was given than was ordered," he says, "so, for example, there's a physical constant in quantum electrodynamics that is known to nine decimal places. How can it be that this brain that we have is able to do maths that can get to that constant and yet this thing was evolved in order to solve problems about how to pitch a spear or how to track something? You don't need nine decimal places to make a calculation of that kind in your head. It doesn't require the kind of accuracy that's going on here."

The evolutionary orthodoxy states that nothing was ordered and nothing was given. The brain merely evolved in response to certain environmental pressures. It evolved, as Kaiser says, to pitch a spear or track an animal. But, in doing so, it attained a level of complexity that allowed it to go far beyond pitching and tracking. Once that happened, Beethoven's late quartets, quantum electrodynamics, and genetics were just a matter of time, and meaning was an illusion, an epiphenomenon.

But the urge to find meaning is strong within all of us and there is, as I said, one possible satisfaction available in the hard scientific world view. Some who believe absolutely in the ultimate competence of science, in the evolutionary orthodoxy, are driven to find human purpose in that. E. O. Wilson does this, insisting that a true perception of our place in the biological scheme of things will provide the meaning we crave and will supersede all previous mythologies. Daniel Dennett writes of the way "*importance itself*, like everything else that we treasure, gradually evolves from nothingness."

Biology explains our need to know and to have purpose. The world seems to be about things; it has what philosophers call intentionality. And, to Dennett and Wilson, it has this property because of life. Purpose came into the world when the first successful system of replication—the precursor of all life—evolved in the primordial swamp. From that point onward a direction existed, the direction of life. Before that moment there were only accident and contingency. Afterward there was purpose. This purpose is hardly sublime; it is merely the determination to survive, to create another generation. But from it flow all other purposes; from it flows meaning. The gene, the atomic unit of the purpose of replication, becomes the "unmeant meaner," the starting point of all meaning. Knowing this, we can abandon the need for any mystery or external justification. The meaningful gene fills the empty shrine of secular society; it replaces the Christian soul. We are what we appear to be; we mean what we appear, under the microscope, to mean. And that should be meaning enough for anybody.

There is a problem with this, a problem of which both Wilson and Dennett, in their different ways, are aware. Of course, if you have faith in a divine purpose, the problem is that they are both simply wrong. But with or without faith, the real problem is that a purely science-based form of meaning may be inadequate. Nature has no obligation to satisfy our craving with the consolation of comfortable meanings. The scientific materialist mythology may not be what we want. The news from the space stations and the laboratories may be bad; it may be inconsistent with human longing. It may thwart our highest aspirations. Knowledge may be intolerable.

This is why I began my book with that piercingly profound quotation from Shakespeare's *The Winter's Tale*. Imagine this, says Leontes, King of Sicilia: you drink your wine from a cup with a spider at the bottom. But you do not see the spider. The wine tastes good. Then somebody shows you the spider. Suddenly the wine is disgusting; you feel sick. It is the same wine, but it has been changed by the intolerable knowledge of the spider. The world is, apparently, unchanged, but knowledge has, in reality, changed it utterly. What if genetic knowledge is the spider in the cup?

Dennett and many others confront this problem by insisting that it is not a spider at all. It is, rather, a special device for improving the taste of wine. The news is, in fact, good; the Darwinian world view is "just what we need in our attempt to preserve and explain the values we cherish." The whole of the new discipline of evolutionary psychology, a discipline that grew out of Wilson's work, attempts to back this up by saying that what goodness, what moral impulses we have, are bequeathed to us by our Darwinian inheritance. The scientific myth replaces the religious as the basis for our lives and values.

But, generally, contemporary technocrats do not see the need for such myth-making. They simply shrug their shoulders and say: We are sorry but this is the truth, make what you can of it. There is no problem because, on the whole, science seems to work.

"Understanding beats not understanding," says Eric Lander, "and I stand by that."

"History speaks strongly on my side," says David Botstein. "We manage as a society to do more good than harm."

For Lander and Botstein the scientific project is working, and behind their words is the implicit insistence that there is, in fact, nothing else to be done. Scientifically investigating the world is what we do; it is the climax of the logotaxis that led us out of the Great Rift Valley. Everything else is a sideshow, recreation, a game. Meaning, in particular, is a distraction.

"I never was interested in the meaning of life as such," Botstein said. "I take that to be a very humanistic position. In a way I think it's a simple view that life has meaning. My own view has always been that we are here to make the best of it. For whatever reason, we do have some sense of what is progress and what is good. My own view was that the one thing that seemed like a proper thing for a young human with any ability at all to do was to try and understand the world better."

Of course, we can continue—we have continued—with our disagreements between those, like Botstein, who feel that science is enough and those, like me, who feel that it isn't. In practical terms it can make very little difference. For example, James Watson observes DNA through the eyes of an atheist; Francis Collins, through the eyes of a Christian. It looks like the same stuff and they both remain central figures in that supreme project of contemporary science—the sequencing of the human genome. The absolute contradiction in their world views does not appear to affect what they do. Collins says God wants us to have this knowledge; Watson says it disproves God by removing the need for Him as an explanatory mechanism. Only when the question is asked does the difference emerge; the rest of the time they get on with the same science and work to the same ends.

But, after Darwin, Mendel, Crick and Watson, after Dolly the sheep, I don't think we can seriously expect to rely on this peaceful coexistence. Biology in general and genetics in particular have taken us too far; they have taken us over the edge.

The new biology is now crossing the final barrier. Previous science may have made it hard, if not impossible, to believe in a God beyond the clouds or heaven above and hell beneath. Its effectiveness may have persuaded us that it is the truth and that it grants us the power to do almost anything. But always we had the ultimate refuge of our selves. There was always something irreducible about our experience and identity, something that lay beyond science, however effective. We may examine the world as scientists, but only because, as human beings, we felt we were not entirely part of it. The trick of self-consciousness gave us a dual sense of ourselves as being both within and beyond the world.

"We feel that even when all possible scientific questions have been answered," wrote the philosopher Ludwig Wittgenstein, "the problems of life remain completely untouched." Such a remark could still, before genetics, seem like the most obvious common sense; however far our science took us, it would never even scratch the surface of "the problems of life." And that was because these problems were not a part of the world that science was so effectively explaining. They belonged to another world, the world of human experience.

But the story I have been telling is the story of how science has come to assault that last refuge. For some this assault is a glorious, climactic battle. Writing of both Darwinism and the study of artificial intelligence, which aspires to make computers think like humans, Dennett says: "Together they strike a fundamental blow at the last refuge to which people have retreated in the face of the Copernican Revolution: the mind as an inner sanctum that science cannot reach."

I must pause in wonder at the barely veiled brutality of such language. Who are these people? Everybody except Dennett? And why did they retreat? Or rather: Why shouldn't they retreat? Is Dennett implying that they—we—are just too cowardly and self-deluding to face the truths of science? If they are, those characteristics must also be a product of the Darwinian evolution which this philosopher is so determined to celebrate as the solution to all our problems. And if that is the case, then Darwinism is both problem and solution, a weird type of theory that can explain anything you might throw at it; as John Vandermeer has written, "A theory that explains everything is sort of like God." The battle that Dennett celebrates is not that of science against superstition; it is, in truth, that of one god against another.

The real point is, of course, that there was a reason for that retreat, a need for that refuge. The reason was that people could not believe that their lives could be accounted for in these terms. The need was for something—in this example the mind—that seemed to stand decisively beyond the competence of such accounts. And why did they need that reason and feel that need? Because they felt incompetent. Like all people at all times in all places, they felt uncertainty and impotence. They mistrusted these powers being claimed for humanity because they saw the mess and indirection of human life. God makes perfect sense to anybody who honestly faces the gulf between the human ability to conceive perfection, goodness, and harmony and the facts of the world. And, if we cannot have God, then we must construct a place beyond the reach of the so often disastrous tinkerings of human reason. That place was the human mind.

Yet, out there in the contemporary world, Dennett wins the argument. It does not matter whether he is right, whether science can or cannot reach the mind. What matters is that science has done so much else. It has convinced us by remaking our world in its own image. It can encompass the stars and dissect the atom; it can transform our lives beyond recognition. Nothing, surely, can be beyond its powers. Secretly we must suspect, even in our last post-Copernican refuge, that our security is temporary. It is only a matter of time before science reveals the walls around our selves to be built of nothing more substantial than consoling fantasies.

Genetics is the knock on the door. James Watson made it clear that this is the ultimate science when he told me, "The end result of the human genome program on society will finally be to make people realize we are the products of evolution, not of a message from the sky. Finally they are going to find it impossible to ignore."

People may have resisted the lesson of science throughout the triumphs of physics, but, after genetics, they will be able to do so no longer. Now we know, says genetics, what you are all about. At the molecular level we have found the core of your being.

"Everything we know about molecular biology," writes Francis Crick, "appears to be explainable in a standard chemical way. We also now appreciate that molecular biology is not a trivial aspect of biological systems. It is at the heart of the matter. Almost all aspects of life are engineered at the molecular level, and, without understanding molecules, we can only have a very sketchy understanding of life itself. All approaches at a higher level are suspect until confirmed at the molecular level."

Crick really meant it when he burst into the Eagle to announce he had found the secret of life. Other geneticists might disagree. Francis Collins makes the common point that our molecular knowledge can only go so far. "It doesn't alarm me personally," he says. "It's not a philosophical assault on what I believe. We've always known that various parameters of humanity are hereditary as long as we've been studying identical twins. It's clear that a lot of aspects of behavior, personality, and illness are encoded. That doesn't shake me up. It doesn't threaten my own belief that probably we won't understand a lot of the human spirit on the basis of DNA analysis. A lot of important things like love will remain wonderful and mysterious when we have those three billion base pairs on the computer."

Love becomes the name of that last refuge. It has been given other names: God, the soul, values, society, history. But, thanks to the corrosive ambitions of genetics, this doesn't work either. Look again at Crick's last sentence: "All approaches at a higher level are suspect until confirmed at the molecular level." All truths are to be tested against the molecular truth. No matter how elaborate, complex, moving, or profound we may find our world, it is, ultimately, no more than a molecular function.

Of course, we could still say, Yes, but it doesn't matter. Either we could simply assert that we don't believe this hard reductionism, or we could say that *ultimately* everything may be molecular, but so what? The chain of connection between the molecules and, say, the Sistine Chapel ceiling is so long that it is meaningless. We can still wonder at that great work of art. Human greatness need not finally be reduced.

This is, in fact, one version of the single most commonly advanced argument and, as it goes to the heart of the matter, I shall pause to deal with it at some length. It is central to the optimism of writers like Dennett and Robert Wright, and, in one form or another, it occurs repeatedly in books and debates about the impact of science on the world. The argument is, essentially, that science does not affect our spiritual capacity because the causal chains are so long and our experiences so complex that, spiritually, we can still feel untouched.

Here, for example, is Dennett on the music of Bach: "Bach is precious not because he had within his brain a magic pearl of genius-stuff, a skyhook, but because he was, or contained, an utterly idiosyncratic structure of cranes, made of cranes, made of cranes, made of cranes." Skyhooks are, in Dennett's terms, spurious, nonmaterial interventions in the world, and cranes are the real, material systems that make things happen.

And the physicist Richard Feynman wrote of how the knowledge of science "only adds to the excitement and mystery and awe of a flower. It only adds. I don't understand how it subtracts." In a slightly different form, the argument is employed by Darwin in the very last sentence of *Origin of Species*: "There is grandeur in this view of life, with its several powers, having been originally breathed into a few forms or into one; and that, whilst this planet has gone cycling on according to the fixed laws of gravity, from so simple a beginning endless forms most beautiful and most wonderful have been, and are being, evolved."

David Botstein makes yet another version of the same point when he insists that the essential human experience itself remains unchanged by explanation: "Having understood the physics of why the sky is blue doesn't make it any less blue."

Feynman's "excitement and mystery and awe," Darwin's "grandeur," Botstein's blue sky, and even Dennett's "utterly idiosyncratic structure" are all ways of saying the same thing. We do not need magic, mystery, or anything extrascientific in order to experience the spiritual. A flower or the music of Bach are both unaffected by the fact that we can fit them into a material, explanatory system. Science may be assaulting the last refuge of the self, but the self need not necessarily worry.

The first point to note about this argument is that it is highly subjective. In fact, it is not really an argument at all. All each of these men is saying is that *my* spiritual experience is untouched by the explanations of science and, therefore, there can be no reason why anybody else's experience should be affected. They remind me of those oppressively upbeat weather forecasters on morning TV shows who seem to be perpetually insisting, "I feel fine so you should feel fine too; in fact, if you don't feel fine, there's something wrong with you." This is a statement of faith, hope, or emotional bullying rather than logic. And, anyway, I don't feel fine and any fool can be a weatherman.

The second point is that, if the last four hundred years of history have taught us nothing else, they have surely taught us that science does affect our spiritual sense very deeply indeed. Most of the philosophy and theology and much of the art of this period have been concerned with the search for a place in the world, for a stable set of values and meanings. That search has been driven by the awareness that modernity, the age of science, had undermined all previous stabilities. If it had not been for the scientific revolution we would not have had the painting of Francis Bacon or the writing of Samuel Beckett or Saul Bellow. These are works built upon the sense of fragmentation and alienation these artists perceived in the modern world view, the scientific world view. It is relatively trivial in this context subjectively to claim that our appreciation of Bach or the Sistine Chapel ceiling is unaffected by our scientific modernity. How, apart from anything else, would we know? The much deeper point is that something in what we have become as a result of the pressure of science on the evolution of art prevents us from composing or painting exactly like that.

All of which is to make what should be an obvious point: The explanations of science emphatically do not leave our spiritual sense—as manifested in our art or our perception of the world—untouched. They change it radically, and I suspect the slight hint of desperation in the concluding words of Darwin—"There is grandeur . . ."—indicates that he, at least, was a man great enough to know this perfectly well.

In this context, Francis Crick's words about all higher-level approaches needing to be confirmed at the molecular level can be seen as a spiritual as well as a scientific statement. In this new form of spirituality we cannot really trust anything as having any final meaning or validity until it has been underwritten by the molecular geneticist. This is, in fact, exactly what sociobiology, evolutionary psychology, and all the other attempts to place the gene at the center of our lives are saying. Although they speak of large-scale effects—human and animal behavior individually or en masse—they do so only in terms of the replicating demands of the gene, in terms of a fragment of a molecule of DNA. Everything, in these interpretations, flows from the logic of that initial replicating drive. These developments spring directly from genetics and they represent the final extension of the powers of science into the most intimate human realm, into our last refuge.

But why can't we just go on? We seem to be doing so after all. In fact, we seem to be enjoying it. The material fruits of science are welcome. They make life easier, richer, and safer. We may have just lived through the bloodiest and most scientific century in history, but perhaps we have now learned from our mistakes. No Lenin or Hitler, armed with new perversions of science, seems to be on the horizon, no new beast, in W. B. Yeats's words, seems to be slouching toward Bethlehem to be born. Perhaps we are entering a new era in which, freed of all our old illusions and armed with our formidable knowledge, we can find peace in the world.

I can find few genuinely deep thinkers who believe this. To be honest, I can find none, for, in truth, the belief itself is intrinsically shallow. History does not work like that and history, in spite of reports to the contrary, is not over.

Let me return to the quotation with which I began this chapter. Isaiah Berlin sees such extraordinarily intricate elaboration in human experience that the very idea of a single "infallible scientific key" amounts to "one of the most grotesque claims ever made by human beings." Roger Scruton, the philosopher, has written "that scientific truth has human illusion as its regular by-product, and that philosophy is our surest weapon in the attempt to rescue truth from this predicament." Both see that, as a moral or spiritual guide, science is utterly inadequate—indeed, largely meaningless.

And yet, today, that is precisely what it aspires to be. That is what Dennett, Wilson, and Wright are saying. A huge division has opened up within the culture between those who say science can save us and those who say it is the one thing that most definitely cannot. It is the division between Edenists and evolutionists with which I started this chapter.

The science side is winning, not just because it has the most money and the most power, but because it has no serious competition. For the further twist of this particular knife is that science is offering its salvation at a time when alternative explanations or ways of understanding seem to have collapsed. These are the words of Howard Kaye, a sociologist: "As our latest attempt at dropping some moral anchor, biology may prove as ambiguous and unsuccessful as previous scientific moralities—and perhaps even more harmful. Our current infatuation with biology, unlike that of a century ago, is occurring at a time when the humanities and social sciences have declared moral bankruptcy, thus depriving us of a vital part of the collective memory we need to regulate and resist our increased capacity for genetic manipulation."

And here is the philosopher Gregory S. Kavka: "This combination of vast power, a lack of clear standards, and inescapable choices is a prescription for a tormented collective soul and, at the operational level, for intense political conflict."

And Timothy F. Murphy points out the extent to which we have accepted the rule of science that we must know everything. We are prepared to indulge this rule without considering the consequences to "our social mores and institutions." In other words, we don't believe in anything as much as we believe in science.

These thinkers—and, I hope, this book—make it clear that we are confronting a new ideology of scientism. This is the belief in the absolute power and competence of science. It is a belief that is often implicit in the writings of scientists but seldom explicit. It is a belief that invalidates what was, until recently, an unspoken treaty between science and the humanities. The terms of this treaty are, basically, that the two are quite separate. The humanities are about values and science is value-neutral. They may both describe the world, but in different, noncontradictory ways.

This is how the British geneticist Steve Jones describes the terms of this treaty: "This means that in science it is not enough to search ancient texts or even modern philosophy for solutions; instead there has to be objective enquiry into rules that apply to all creatures, human or not. As ethical beliefs are themselves a human construct they cannot be studied in this way. In spite of the hopes of those searching for a salvation in modern biology, the double helix has no moral content."

The great British scientist Peter Medawar argued that this treaty was not merely an agreement, but was, in fact, logically necessary. His Law of the Conservation of Information was, essentially, that you couldn't think your way to things that weren't there in the first place. "The Law . . . ," he wrote, "makes it clear that from observation statements or descriptive laws having only empirical furniture there is no process of reasoning by which we may derive theorems having to do with first and last things; it is no more easily possible to derive such theorems from the hypotheses and observation statements with which science begins than it is possible to deduce from the axioms and postulates of Euclid a theorem to do with how to cook an omelet or bake a cake—accomplishments that would at once unseat the Law of Conservation of Information. I do not believe that revelation is a source of information, though I acknowledge that it is widely believed to be so. . . ."

But scientism accepts neither the treaty nor the logic. It demands total power. Everything can be explained and understood scientifically. This exclusive realm claimed by the humanities or religion is an illusion, a rhetorical artifact. In time it will be annexed by science. This is precisely what genetics—explicitly—is now doing. It is claiming to have found the basis of all human experience, values included, and so the treaty is annulled and the scientific society is born. Jones, in these terms, is wrong; the double helix is, in fact, replete with moral content because it is where morals start.

For Berlin, this idea is an illusion and, for Scruton, it breeds illusions. For Kaye and Kavka, it is threatening, a development with which the human race is utterly unable to cope. For Murphy, it has already destroyed our power to protect our social institutions. Science, for all these thinkers, should only ever be a part of the picture. If it becomes the whole picture, catastrophe ensues. For we must be able to judge and evaluate science; it must be a part of a culture. If we can't, there is nothing to discuss. Whatever science and the scientists say must be right. Step over this cliff, they might say, the fall will do you good, and we can only believe they must be right.

Of course, we can set up our ethics committees and President Clinton can demand an ethics report on the implications of Dolly the sheep. But, in reality, these ponderings and reports only provide snapshots of the intensity of the residual resistance to scientism within society. Such devices ask merely, What is left of our ethics that we must, for the moment, place in the way of the scientists to show that we are doing something meaningful? The real question should be: Are ethics possible if genetics explains everything, including our values? How are we to evaluate this ultimate theory? What logic, what language can we use if, for example, we wish to ban human cloning? How are we to legitimize our fear and disgust? A theory that explains everything is, indeed, like God, and, like God, it will not be mocked.

The truth is that we have no language with which to mock this new god. This is why the optimism of so many science writers rings so hollow and why libertarian free marketeers sound so shallow on the subject of science. One after another they assume that, somehow, we—or the market—will do the right thing. But they do not realize that their definition of the "right thing" is derived from a complex cultural legacy of which science is only one small part. They assume this legacy is intact to guide us while, in the same breath, they celebrate its extermination by the ideology of scientism.

This point has always seemed so obvious to me that I was alarmed that only I appeared to be aware of it. So it was with some relief that I found it expressed in an essay by the historian Mitchell G. Ash. He discusses the usual, crudely libertarian claim that the best way of handling genetic technologies is to leave them to the market, allowing consumers freedom of choice: "The concept of free choice has become part of America's civic religion. Those who proclaim a faith in 'patient autonomy' and believe that the good intentions of physicians in this regard are enough to ensure 'good' choices simply miss the point. For after all, where did the values and beliefs that govern those choices come from?"

Exactly. The values and belief by which we are supposedly judging the offerings of genetics are themselves the product of a scientific society. Our apparently free choice is always rigged. We have simply lost the ability to think of a coherent reason not to accept whatever gifts the technocrats choose to bestow.

The language we are able to use changes the moral context in which we are obliged to live. So we could not now respectably speak of "the improvement of the race" or of "selective breeding"—the terminology of the old eugenics—but we do speak of the "quality of life" and assess our children in consumerist terms. Only the names have changed. The now acceptable virtues of the consumer society replace the now unacceptable ones of eugenics and fascism. Beneath lies the same impulse—to bring the most intimate aspect of human life under the control of a scientistic ideology.

So the problem I am posing is this: On one side is a supremely successful and effective way of knowing the world and now, thanks to genetics, ourselves; on the other is the disparate and, compared with science, relatively ineffective debris of our nonscientific culture. The latter is being invaded by the former. Aware of this, scientists like E. O. Wilson and philosophers like Daniel Dennett argue that the invasion is, or at least can be, benign and that we can derive new values from the insights of science. Less ambitiously, scientists like Eric Lander and David Botstein insist simply that it is what we humans do; it makes sense and it seems to work. Or there are people like Berlin, Scruton, and, to place myself briefly in this exalted company, me, who do not believe that any such values are possible or meaningful, but rather that they are immensely dangerous. We may be aware that nonscientific culture presents a messy spectacle of wreckage, dispute, and indirection. We may fear that, without the unifying force of a religion, rebuilding anything from this wreckage may be difficult if not impossible, but we are convinced that it is necessary if we are not to slip into a new dark age of illusion in which we become passive, manipulated consumers of whatever science and its eager paymasters have to offer.

So how do we regain a stable, nonscientific sense of ourselves? The first answer is that science—not religion, not mysticism, not astrology, not belief, nor any transcendent moral claims—should now be the proper object of our skepticism. Science should, at every turn, be restrained by our doubt. Even Francis Bacon, the sixteenth-century philosopher of the new age of science who made the most ambitious claims for human omniscience, was aware of this need for restraint. He wrote of three limitations on the power of science: "The first, that we not so place our felicity in knowledge as we forget our mortality: the second, that we make application of knowledge to give ourselves repose and contentment and not distaste or repining: the third, that we do not presume by the contemplations of nature to attain the mysteries of God."

In other words: Be humble—one lesson the current wave of propagandists for science, most of whom would unthinkingly canonize Bacon, tend to forget. And we, above all people, should know the practical and historical reason for this humility—the long catalogue of errors associated with science. Science has failed to make sense of economics; it failed, almost catastrophically, to analyze global strategy effectively during the Cold War; it repeatedly fails—as in the use of DDT as an insecticide and in numerous environmental areas—to foresee the full impact of its technology; it failed in the '60s and '70s to find what was thought to be the viral basis of cancer and so on and so on. The list is endless.

And it is a list created by arrogance. For notice the strange way that, in every age since its creation, modern science has always seen itself, and been accepted, as right. For some reason, when it moves into the realm of public policy, science always forgets its own essential nature as a provisional system of knowledge and begins to make outrageous claims.

Remember that, in almost all areas, scientific truth is a moving target. Einstein revealed that Newton was only provisionally the Truth. What we believe now about the nature of the world is utterly different from what was believed a hundred years ago, and, if we go back two hundred years, it would be quite unrecognizable. Stephen Hawking, that great icon of the heroism of modern science, has changed his mind fundamentally within his own working lifetime. Can we really be sure that our current scientific beliefs will be intact, even in part, in another two hundred years? They might be seen as wrong, partial, inadequate, or even just as mad as Aristotle's physics now seem. And, if there is so much doubt, how can we possibly sacrifice so much of what we are to a few tentative ideas that, we should know from history, are likely to be proved wrong? It is always difficult to see around the beliefs of one's own time; it may seem logically impossible. But plainly, if these beliefs are claiming as much as they now are, it is necessary.

So the first moral of this book is: Since science is the most powerful and wealthiest orthodoxy the world has ever known, it must therefore be the target of the most rigorous skepticism.

The second way to rebuild is to keep in our mind all the time that science is so obviously not enough. In many ways this remains deeply embedded within our culture. Popular science fiction, a form that is driven overwhelmingly by the thrills of new or alien technologies, is also replete with carefully created, nondenominational, politically correct spirituality. In the film *Star Wars*, for example, there is the Force, a generalized, universal power with good and bad sides which confers power higher than that of technology. "It's an energy field," explains Obi-Wan Kenobi, "created by all living things; it surrounds us and penetrates us; it binds the galaxy together." When Luke Skywalker attacks the Death Star, he turns off his targeting computer to rely instead on the Force. In both the films *ET* and *Close Encounters of the Third Kind*, the stories are based on the spiritual experience of a contradiction between the sublimity of the alien contact and the clumsy, brutal technology of our world. And *Star Trek*, in film and on television, is specifically intended to be one long spiritual journey.

It is easy to mock such examples, easy to say they mean nothing. But, in fact, if you look at these hugely popular fairy tales carefully, it becomes clear that they simply could not work without this spiritual dimension. Other, less technologically dependent fairy tales can. But when the technology takes over, it seems to be necessary to create a transcendent space into which we can escape. For, of course, without it there could be no drama, no involvement, only the conflict of rival technologies. Without the Force, *Star Wars* would not be any kind of recognizably human story at all.

Such popular myths are evidence of the way extrascientific ideas seem to be necessary for the conduct of any human life. It is not just that we like to believe in such things, that we are inveterate wishful thinkers, but rather that we cannot function without them. Even the most skeptical scientists explicitly or implicitly acknowledge this in a variety of odd, often casual, but nevertheless revealing ways. E. O. Wilson begins his beautifully written book, *The Diversity of Life*, with an evocation of the Amazon rain forest at night: "The unknown and prodigious are drugs to the scientific imagination, stirring insatiable hunger with a single taste. In our hearts we hope we will never discover everything. We pray there will always be a world like this one at whose edge I sat in darkness. The rain forest in its richness is one of the last repositories on earth of that timeless dream."

"Would you want total knowledge?" I asked James Watson.

"No," he replied. "I'd rather play tennis."

A different though closely related theme emerges in passing remarks like this one from the geneticist William Cookson, who is considering the possibility of parents being able easily to select the sex of their child: "This horrifies me," he writes, "although I cannot say why."

Peter Goodfellow repeatedly referred to this kind of indefinable unease in our conversation, and something similar has cropped up over and over again in my conversations with scientists and my reading of their work.

I take this unease to be evidence of something scientists find difficult to acknowledge or analyze. I too find it difficult to explain, but it has, I suppose, something to do with the nagging subjectivity that seems to lie behind their enterprise. They are, after all, only human; they have, therefore, doubts even when considering the apparent certainties of the scientific enterprise. On the one hand, Watson casually and Wilson passionately seem to resist the possibility of the completion of the scientific task. On the other hand, Cookson and others find themselves with doubts about where their work is leading, doubts that seem irrational and beyond analysis. Yet completion is an ambition intrinsic to the scientific effort. Completion would dispel any remaining doubts about the competence of science. And anxieties about where science is leading are irresponsible to those who believe in it as a moral force and as the defining human project.

So there is a conflict—acknowledged by the best, concealed by the worst—between the private minds of the scientists and the public propaganda of science. Privately they know their subject does not provide the ready, simple answers they must publicly claim to get funding or to sell books. Rather, it confuses them as much as it confuses us.

The reason for this conclusion is that, at least for the moment, nobody is capable of finally handing himself over to the creed of scientism. There is still so much within us—whether it is called God, tradition, culture, conscience, or history—that seems to refer to or to be based on an extrascientific realm. Ethics committees often pay lip service to this when they attempt to find some way of assessing social attitudes to cope with some new development. But they tend to do so feebly, probably because the huge plurality of the systems that might be used to justify such qualms—Christianity, Islam, Buddhism, humanism, capitalism—makes it difficult to be precise about policy recommendations. In effect, the qualms are reluctantly accepted, but only on the basis that they are little more than the residues of ancient superstitions. Perhaps, runs the unwritten subtext of all those ethics reports, science is not enough for those of us alive now, but for future generations, freed of these mental relics, it will be and the ethics committees can finally be disbanded.

But why should we wish to abandon our nonscientific selves? If, as I said earlier, science is repeatedly wrong and needs judging, then surely these are precisely the selves we need most. We may find it hard to define exactly what we are talking about. But of course it is hard, and it always has been, to resist the dominant power of the age. Look at the lives of Galileo, Solzhenitsyn, and George Washington.

Most often resistance comes in the form of simple, reflex revulsion. In the British coverage of the new genetics there has been much talk of the "yuk" factor—the involuntary disgust felt by many to specific activities of biologists. I think the "yuk" factor first appeared in response to the idea of using the eggs from aborted female embryos to help infertile couples. Many argued that "yuk" means nothing, it is merely an unthinking response which has not fully comprehended the science or taken in all the arguments. Well, it is certainly true that the "yuk" reflex passes. It may have been the first response to the idea of surgery or organ transplantation, but these practices do not now inspire many "yuks."

Yet "yuk" may mean a great deal. It obviously has much in common with Cookson's "This horrifies me, although I cannot say why." And obviously, in its spontaneity, it is an expression of the idea that some barrier seems to have been crossed, some taboo violated. Clearly, different people will say "yuk" to different things. And when it comes to new genetic manipulations, the expression of "yuks" are likely to be equally culturally diverse.

But that does not mean "yuk" is an imprecise measure of moral resistance. It means exactly the opposite. For it says that, whatever the occasion of the offense, people are still capable of being offended. People habitually, though often unconsciously, establish taboos—lines of the sacred that must not be crossed. When those lines are crossed, people inevitably respond. It is not for scientists or anybody else to tell them they are wrong. They would be better employed understanding how such reflexes may well have sustained—may well still be sustaining— human societies. The human world can, after all, be seen as a system of lines, of restrictions and taboos, a systems of "yuks." How else could it possibly work?

This is related, of course, to the point I made earlier about people's awareness of human incompetence and impotence. A realm of the taboo or sacred is precisely intended to be a place beyond the hamfisted interventions of human beings. "Yuk" means we are not supposed to be here, doing this; we are not good enough or clever enough. This need not be a religious response. The theologian John P. Boyle, in discussing the possibility of germ-line intervention, remarks: "Even those who do not believe in God may find themselves opposed to 'playing God' with the human genetic endowment." And Boyle adds: "Among such people there is, I suspect, a reluctance to tamper with a given in our nature even when the intervention is planned for reasons that are therapeutic and not eugenic." The philosopher Leonard M. Fleck has pointed out that traditional theories of justice are radically undermined by genetics. They "operate in a world in which natural assets and liabilities are assumed as given." If genetics allows us to change people, then it redefines our sense of what people are and casts doubt on the meaning and viability of our most important institutions.

Lines are important. And one of the most important lines Western culture has drawn is around the human individual, not as the crass economic absolute of the free market fundamentalists, but as the moral absolute of Immanuel Kant. As such, the individual represents something that is "given," and this idea of a "given in our nature"—however metaphysical—suggests a distinct realm which it would be dangerous for us to invade. That seems to me to be a valuable, indeed a universal, idea that expresses a human truth and demands, from geneticists in particular, a very cautious and humble approach.

One final point needs to be made under my heading Science Is Not Enough. Up to this point I have been saying that science is not enough because of the way we are. Clearly, however, we could be different and then science would be enough. We could be changed, "improved," so that we would be more ready to accept the reign of science. This may sound insanely dictatorial. It is. But such an idea is, in fact, implicit in much of what scientists say. Both Francis Crick and James Watson, for example, believe that part of their task is to rid people of illusions so that they may fully accept the truths of science. The project is to make us undeluded citizens of a scientific world. The same idea was present in the suggestion by the physicist Steven Weinberg that once physics had its Theory of Everything, people would stop reading horoscopes. Scientific truth would eliminate the need for all competing truths. Finally, we shall be free of the deluded debris of the past.

The sheer arrogance of this idea seems to escape the notice of these scientists. Plainly they have now comprehensively torn up the treaty between science and the humanities. What they do has become the only thing to be done. These are, truly, grotesque claims.

Could it work? I do not believe so, but I am unable to prove it. There may be no logical reason why a fully scientific society is not possible; it merely seems an incredible and highly unpleasant idea. Indeed, it is so unpleasant that even these propagandists of science do not quite say that is what they want. It would be a huge rhetorical blunder, alienating their audience. Imagine saying, "I want everybody to accept everything we scientists say," on the Oprah Winfrey show. You would rightly be laughed at. Nevertheless, that is what is actually being said.

But the reason I believe a fully scientific society could not work is that there does seem to be a profound logical discontinuity between the insights of science and the way we actually conduct our lives. Human society is *intrinsically* unscientific. The best way of explaining this is through the concept of free will.

Whether or not we have such a thing as free will—a central element in Christian theology as it is a way of explaining why and how evil entered God's creation—has been a prolonged philosophical and, to some extent, scientific debate. Scientifically, truly free will is improbable if not impossible. The world is a structure of cause and effect: one event causes another. If we knew every event down to the subatomic level, then we could know every other. We could see that everything was determined and I would realize that I am deluded in my belief that raising my right arm at this moment is a freely chosen act. In biology, as usual, this becomes more personal. This is a vision that was expressed most dramatically by Pierre Simon de Laplace, the mathematician who told Napoleon that he had no need of the God hyothesis. "An intellect," he wrote, "which at any given moment knew all the forces that animate Nature and the mutual positions of the beings that comprise it, if this intellect were vast enough to submit its data to analysis, could condense into a single formula the movement of the greatest bodies of the universe and that of the lightest atom: for such an intellect nothing could be uncertain; and the future just like the past would be present before its eyes."

Genetics takes this possibility into the human realm.

"The ontological implication of human genetics," writes the philosopher Michael Ruse, "is that there is no such *thing* as free will. You do not have the body produced by the genes and then free will on top. This is not to say that humans are not free, but rather that freedom must be sought within the bounds of ontological reductionism."

Ruse's point is that free will is not some separate, metaphysical entity, but rather a part of the biological mechanism. We are not free of the deterministic prison of cause and effect—ultimately we must be in that prison—but we are, somehow, wired up to act and believe that our will is free. It is a kind of necessary fiction.

The odd thing about free will versus determinism, however, as the philosopher Panayot Butcharov has pointed out, is that it makes little difference which we choose, "because any theory about it is compatible with whatever choices one in fact makes." I can equally well fit my raised arm into a theory of hard determinism or one of total free will. Nothing is changed by my interpretation. But the point is, we always act *as if* we were free, and this raises the question: Is it even possible to act *as if* we weren't?

As Butcharov puts it: "What matters is whether a person possessing such a belief would almost inevitably think of himself/herself in a hard-deterministic way, and, more fundamentally, what it would be like to think of oneself in such a way. As Kant and many others have remarked, everyone who is engaged in thinking or choosing, or making a decision presupposes that this thinking, choosing, or deciding are not causally determined in advance. We do not have a clear idea of what it would be for one to have *internalized* a fully deterministic conception of oneself."

In short, it appears to be impossible to act at any given moment as if we weren't free. This is not a remote philosophical point. Consider, for example, its implications in the realm of justice. We treat people as innocent or guilty because we need to judge them *as if* they were free. Clearly we modify this with the concept of mitigating circumstances, but such a modification must be placed on top of a basic idea of freedom; it does not change things fundamentally.

So here is a scientific truth in clear and unresolvable conflict with a human truth. We simply cannot incorporate Laplacean determinism into the way we live. Just being human is a task that requires extra-scientific wisdom. It is absolutely impossible to incorporate all the insights of science into a functioning world view. What we think we know is implacably opposed to what we think we are. But what will happen if we try to forcibly unite the two?

"Ultimately," write Joseph Levine and David Suzuki, "the impact of molecular biology on the study of behavior will be determined by our ability to face the prospects that the reductionist vision may become reality, that someday we may be able to determine, sequence, and develop tests for genes that powerfully influence the development of our intelligence and other complex traits."

Do we have any such ability to face such prospects? If we do not, and I don't think we do, then science is emphatically not enough, and, in pursuing the idea that it is, we are embarked on a course of self-destruction. In order to become scientific, we must become inhuman. The problems arising from this conflict are with us already—notably in the political realm of equality. We may just have found that an assumption of human equality is the best way to organize society, but biology tells us it is false.

"One of our ideological preconceptions in this country," says Robert Weinberg of the Whitehead Institute in Cambridge, Massachusetts, "is that all men are created equal. That cannot, from the point of view of biologists, be the case. Not only will we find that people are genetically very different from one another, as we obviously must intuit by now, we will find that certain groups, ethnic groups, will be more or less well endowed with certain kinds of genes. It has to be the case from the point of view of the biologist. From the point of view of a democratic ideologist, I am horrified by the notion. But that is the future, inescapably, and we have to deal with it at present."

What we have to deal with is that what is true may not be what works or what we want. There may be a spider in the cup. What we discover may not be neutral, scientific fact, but something deeply offensive to or logically at odds with the way we wish to live. Coming to terms with that possibility must mean that we need a coherent, functioning value system that is independent of any scientific validation. We need to believe in something before we become scientists and certainly before we attempt to apply its discoveries. Otherwise there is nothing to hold us back from the criminal excesses of communism, fascism, or any other malign, scientifically justified ism that may be waiting in the wings to exploit our unbelief.

But, I admit, this is to do no more than establish the problem. Saying we need to believe in something is like saying we need values. Believe in what? Which values?

Today we deliberately construct our societies to avoid answering such questions. Plural, liberal democracies are intended to be neutral systems in which you can believe and value whatever you like within the rule of law. This model has proved immensely successful. Indeed, it represents, according to Francis Fukuyama, "the end of history." There is no longer any serious rival to the liberal democratic model, so history—in the sense of great conflicts between different systems—is over. In the long term this may or may not be right, but in the short term, it seems undeniably accurate. The system works, generating unprecedented wealth and, on the large scale at least, peace.

But this leaves us with a necessary silence on what we need to believe. When confronted with an external threat—Nazism or Communism—the liberal democratic system does seem to acquire a rhetoric of belief. But, without such a threat, there is little that can be said. A moral threat carries nothing like the same weight as the possibility of military defeat. As a result, it seems only logical to fall back on the view that people should make their own decisions. Nondirective counseling and a free market in genetic technologies become the answers because they require no general statement of principles, only the careful cultivation of moral indifference.

This puts science and technology in control. Denied the possibility of general truths, we fall back on the ideology of change and progress. The answer to life's problems becomes a seeking of the solution, the quick fix, that will materially change our situation. We pursue betterment through material progress. But of what does this betterment consist?

To design an angel, it is said, you need a map of heaven. You need to know the nature of goodness and perfection. Well, we think we know how to design angels, but we have no map of heaven. We think we can improve people, but we can't define improvement. So what are we to do now that angel design is widely available? We can talk of enhancement of the human race by the genetic elimination of disease, by choosing the babies we want, or by improving behavioral or intellectual traits. But we can do so without any clear idea of what an improvement would be. Perhaps it is better to suppress aggression, but perhaps aggression is a desirable quality in a businessman or an artist. Perhaps it is better to suppress homosexuality, but was Michelangelo's homosexuality essential to the painting of the Sistine Chapel? Almost certainly.

A conflict arises between what we, privately, may want and what society, publicly, needs. If we as a society need aggression, then should we have the right to stop you as an individual from having an aggressive baby? But society can barely argue about such things; it cannot find the language. The only available language is that of technology: if it can be done, then people should be free to do it, and, in an absolutely plural world, it will, therefore, be done. Say the United States bans human cloning. The churches demand the ban and the people, unbelievers and believers alike, find cloning repellent. But perhaps in Indonesia, Taiwan, or Europe, they feel no such qualms. Americans who wish to clone themselves simply travel. A free global marketplace denies the state the ability to impose its local ethics. Over time plurality produces singularity—a unitary global submission to whatever fruits the tree of technology may bear.

So, heavenless, we design our own private, globalized angels. And, oddly, they turn out to be remarkably similar. For the easiest angel to design is one that is "normal," and normality, in this context, is the global ideal of a person—smart, heterosexual, conformist, and not in-clined to ask awkward questions. Huxley's *Brave New World* suddenly seems not so much a planned utopia, but rather the inevitable product of freedom.

In the novel, the controller, Mustapha Mond, is asked why the peo-ple cannot have both the technological utopia and Shakespeare's tragedy *Othello*. He replies:

> "Because our world is not the same as Othello's world. You can't make flivvers without steel—and you can't make tragedies without social instability. The world's stable now. People are happy; they get what they want, and they never want what they can't get. They're well off; they're safe; they're never ill; they're not afraid of death; they're blissfully ignorant of passion and old age; they're plagued with no mothers or fathers; they've got no wives, or children, or lovers to feel strongly about; they're so conditioned that they practically can't help behaving as they ought to behave. And if anything should go wrong, there's *soma*. Which you go and chuck out of the window in the name of liberty, Mr. Savage. *Liberty!*" He laughed. "Ex-pecting Deltas to know what liberty is! And now expecting them to understand *Othello*. My good boy!"

But what if liberty was the force that created the brave new world in the first place? What if, for example, the liberty to pursue happiness produced this world in which passion and *Othello* were unacceptable?

This may seem a remote, even an absurd, idea. The distance from our real world to Huxley's fictional one is huge. And we know enough about human mess and imprecision to know that we are unlikely ever to create a world as neat and exact as the one in which Mustapha Mond takes such pride. But the worth of such prophetic fiction is not to say exactly what will happen in the future. Rather it is to dramatize the terms of our present predicament. Those terms are, on the one hand, the world as we value it now, with its art and its passions, and, on the other, the world as we might want it to be—easier, less complex, less fraught with suffering. And, in considering these terms, we have to face the fact that Mond is right: Why should we want *Othello*? That play is an expression of unhappiness, a great and beautiful expression, but surely an unnecessary one if certain problems could be solved.

Of course, that is not precisely the issue. Nobody is offering us quite that choice. But they are offering versions of that choice, versions that edge us in the direction of making the decision that we don't, in fact, need *Othello*. The whole brave new world package is not yet being forced on us, and, with luck, it never will be. But, in a thousand different ways, we are being offered fragments of that package.

And the offer comes from a powerful establishment with its own corrupt, consumerist idea of freedom and individualism. It is an establishment that is very fond of genetics. As Philip Kitcher has written, "The ideology of individualism is a powerful friend to genetic determinism. Genetic solutions reassure the successful." And Evelyn Fox Keller, another brilliant, lucid anatomizer of our predicament, has shown the way biology has come to reinforce a particularly convenient view of the human body. "Today's biological organism," she has written, "bears little resemblance to the traditionally maternal guarantor of vital integrity, the source of nurture and sustenance; it is no longer even the passive material substrata of classical genetics. The body of modern biology, like the DNA molecule—and also like the modern corporate or political body—has become just another part of an informational network, now machine, now message, always ready for exchange, each for the other."

Our bodies and our minds—which, of course, in our post-Cartesian world cannot be separated—have been annexed and redefined by science and consumerism. Nobody asked us, nobody gave us a choice. It just happened.

So how do we survive spiritually intact? Some have faith that we just will. The writer Tom Cahill, a friend of mine, has spoken of the world as divided into Romans and Saints. The Romans are the powerful makers and doers, but the Saints are the quiet ones who really change things. Another writer, Colin Tudge, makes a similar distinction between doves and hawks and comes down firmly on the side of the doves.

"In short," he has written, "the science is wonderful and the new technologies are astonishing, but if we are to deploy them for food, we also have to dig deep, and go on digging, down to the roots of our own psyche. Only when we are straight in our own heads, and have structured societies that are able to override their own innate tendency to be overtaken by hawks and hawkishness can we hope to create the kind of world that can be sustained, for only the meek can inherit the Earth."

The survivors will be those who can maintain a sense of spiritual depth within themselves. They will do this in spite of the prevailing ideology which tells them there is no spirit and no depth or which reduces both to a scientifically determined creation story. They will know that the great claims of science foster illusion. And they will, when necessary, laugh at the pretensions of the geneticists when they claim they have discovered who we are. In doing so they will be defending what is sacred, however they may define that troublesome and now beleaguered term.

Whenever I make the kind of points I have made in this book, I am always asked, Yes, but what should we do? I think the question implies agreement in principle to the general ideas combined with a feeling of impotence—a form of fatalism. That is the way the world is going, people are saying; there is nothing that can be done, however much we may dislike it. One thing, however, can done. Minds can be changed and imaginations stimulated to resist the cozy shallowness of the popular scientism now on offer. I cannot tell you how to resist; I can only tell you that you should.

Why? Because genetics asks fundamental questions and offers convincing answers. We live in an age when two big quasi-scientific systems—Nazism and Communism—have been tried and failed catastrophically. We may now be in the process of, unconsciously, trying another, a system based on global, consumerist, technocratic forces. These forces are powerful and transcend the efforts of any individual. There seems to be no alternative to what they offer. We are caught in a trap. Perhaps we always were, perhaps we always will be. But the best are always those who struggle to free themselves from its jaws rather than those who passively accept its restrictions.

I have met and read a few fools and many wise people in the course of working on this book. The fools are all the same, the wise all different. But the wise have something in common: I have no hesitation in saying that the wisest are those with the most doubts, who resist the bland optimism of the propagandists of scientism, and who admit to at least one small, nagging suspicion that not all will be well in the technocratic future.

That suspicion is enough to make my case. Genetics must be contained, humbled—subjected to the freely expressed consciences of human beings who still have a history and who still know the meaning of the spiritual search. And I am consoled, though perhaps not finally convinced, by that supremely religious belief that every honest searcher will, in time, be a finder.

notes

Introduction: The Secret of Life?

1 *Francis winged:* James Watson, *The Double Helix: A Personal Account of the Discovery of the Structure of DNA* (London: Weidenfeld & Nicolson, 1968), p. 197.
5 *"social mores change rapidly":* Jerry E. Bishop and Michael Waldholz, *Genome: The Story of Our Astonishing Attempt to Map All the Genes in the Human Body* (New York: Simon and Schuster, 1991), p. 20.
6 *"There can be no turning back":* Carl F. Cranor, ed., *Are Genes Us? The Social Consequences of the New Genetics* (New Brunswick, N.J.: Rutgers University Press, 1994), p. 43.

Chapter One: The Future

15 *"For almost every behavioral trait":* Joseph Levine and David Suzuki, *The Secret of Life: Redesigning the Living World* (Boston: WGBH Educational Foundation, 1993), p. 240.
18 *"Mice don't get Alzheimer's":* Ibid., p. 188.
22 *"The present state of the brain sciences":* Francis Crick, *What Mad Pursuit: A Personal View of Scientific Discovery* (London: Penguin Books, 1990), p. 163.
24 *the possibility of a Virginia creeper:* Colin Tudge, *The Engineer in the Garden. Genes and Genetics: From the Idea of Heredity to the Creation of Life* (London: Jonathan Cape, 1993), p. 248.
24 *"So perhaps in a hundred years":* Ibid., p. 348.
24 *"All this may mean":* Steve Jones, *The Language of the Genes: Biology, History and the Evolutionary Future* (London: Flamingo, 1994), p. 237.
24 *"The rural landscape":* Ibid., pp. 273–74.

Chapter Two: God, the Bomb, and the Double Helix

26 *For the first time:* Daniel J. Kevles and Leroy Hood, eds., *The Code of Codes: Scientific and Social Issues in the Human Genome Project* (Cambridge, Mass., and London: Harvard University Press, 1993), p. 18.

26 *We believe this issue:* Statement of Dr. Richard Land, executive director of the Christian Life Commission of the Southern Baptist Convention, at the National Press Club, Washington, D.C., May 18, 1995.

27 *"The majority in the scientific community":* Michael Ruse, ed., *But Is It Science? The Philosophical Question in the Creation/Evolution Controversy* (New York: Prometheus Books, 1996), p. 281.

28 *The prospect is that in the next few years:* "Probing Heredity's Secrets," editorial, *The New York Times*, September 12, 1963. Quoted in Carl F. Cranor, ed., *Are Genes Us? The Social Consequences of the New Genetics* (New Brunswick, N.J.: Rutgers University Press, 1994), p. 34.

28 *"The symbol of the mushroom cloud":* Ibid., p. 37.

29 *"For too long the public has maintained a false stereotype":* Gordon Rattray Taylor, *The Biological Time Bomb* (London: Thames and Hudson, 1968), p. 9.

29 *"in life span and the whole pattern of life":* Joshua Lederberg, *Bulletin of the Atomic Scientists* (1966). Quoted in Rattray Taylor, *The Biological Time Bomb*, p. 10.

29 *"The development of biology is going to destroy":* Ibid.

30 *"When man becomes capable of instructing his own cells":* Marshall Nirenberg, "Will Society Be Prepared?," editorial, *Science* 157 (1967), p. 633. Quoted in Cranor, *Are Genes Us?*, p. 36.

30 *"a technology whose consequences":* Leon R. Kass, "Genetic Tampering," letter, *Washington Post*, November 3, 1967. Quoted in Cranor, *Are Genes Us?*, p. 36.

30 *"The more we think about it":* *The New York Times*, November 30, 1969. Quoted in Larry Thompson, *Correcting the Code: Inventing the Genetic Cure for the Human Body* (New York: Simon and Schuster, 1994), p. 54.

32 *"Although I think it's obvious":* Auditotape of the International Conference on Recombinant DNA Molecules (Asilomar Conference), February 24, 1975, Recombinant DNA History Collection, MIT Institute Archives, Cambridge, Mass. Quoted in Cranor, *Are Genes Us?*, p. 42.

32 *"We believe that perhaps the greatest potential":* David Suzuki and Peter Knudtson, *Genethics: The Clash Between the New Genetics and Human Values* (Cambridge, Mass.: Harvard University Press, 1990), p. 214.

33 *"It was quite unprecedented":* Jerome Ravetz, *The Merger of Power with Knowledge* (Herndon, Va.: Mansell, 1990). Quoted in the British Medical

Association, *Our Genetic Future: The Science and Ethics of Genetic Technology* (Oxford: Oxford University Press, 1992), p. 138.

34 *"Have we the right"*: Erwin Chargaff, "On the Dangers of Genetic Meddling," *Science* 192 (1976), p. 938. Quoted in Thompson, *Correcting the Code*, p. 148.

35 *"We Shall Not Be Cloned"*: Ibid., p. 153.

39 *"We, the undersigned religious leaders"*: Land, National Press Club, 1995.

41 *While the contributors to this book disagreed somewhat:* Bernard D. Davis, ed., *The Genetic Revolution: Scientific Prospects and Public Perceptions* (Baltimore and London: Johns Hopkins University Press, 1992), p. 7.

41 *"Is it fair to ask whether anyone"*: London: *The Daily Telegraph* (December 13, 1995).

Chapter Three: Eugenics 1: The Right to Be Unhappy

44 *"In fact," said Mustapha Mond:* Aldous Huxley, *Brave New World* (1932; reprint, London: Flamingo, 1994), p. 219.

46 *"James Watson has articulated a widespread view"*: Robert F. Weir, Susan C. Lawrence, and Evans Fales, eds., *Genes and Human Self-Knowledge: Historical and Philosophical Reflections on Modern Genetics* (Iowa City: University of Iowa Press, 1994), p. 166.

51 *Stephen Jay Gould has written:* Stephen Jay Gould, "Nonmoral Nature," *Hen's Teeth and Horse's Toes* (London: Penguin Books, 1990), p. 35.

53 *I propose to show in this book:* Francis Galton, *Hereditary Genius: an Inquiry into its Laws and Consequences* (London: Macmillan, 1869), p. 1.

59 *"Being cowards, we defeat natural selection"*: George Bernard Shaw, "Preface to Man and Superman," in *The Complete Prefaces of George Bernard Shaw* (London: Paul Hamlyn, 1965), p. 159. Quoted in Daniel J. Kevles, *In the Name of Eugenics: Genetics and the Use of Human Heredity* (Cambridge, Mass.: Harvard University Press, 1995), p. 86.

59 *"The idea of allowing science to interfere"*: Bertrand Russell, "Eugenics," in *Marriage and Morals* (London: George Allen & Unwin, 1929), pp. 272–73. Quoted in Cranor, *Are Genes Us?*, p. 149.

61 *"Society must protect itself"*: Steve Jones, *The Language of the Genes: History and the Evolutionary Future* (London: Flamingo, 1994), p. 282.

62 *"But I don't want comfort. . . ."*: Aldous Huxley, *Brave New World*, p. 219.

63 *A Geneticist's Manifesto:* Hermann Muller, *Out of the Night*, p. 112. "Social Biology and Population Improvement" (the Geneticist's Manifesto), *Nature* 144 (September 16, 1939), pp. 521–22. Quoted in Kevles, *In the Name of Eugenics*, p. 184.

63 *"hailed as an insult to some God"*: J. B. S. Haldane, *Daedalus, or Science and the Future* (New York: E. P. Dutton, 1924), p. 44. Quoted in Kevles, *In the Name of Eugenics*, p. 185.

Chapter Four: Eugenics 2: Tattooing Foreheads

65 *Consume my heart away:* W. B. Yeats, *Sailing to Byzantium*, in *The Collected Poems of W. B. Yeats* (London: Macmillan, 1979), p. 218.

66 *"The artificial selection practiced in our civilized states"*: Ernst Haeckel, *The History of Creation: Or the Development of the Earth and Its Inhabitants by the Actions of Natural Causes. A Popular Exposition of the Doctrine of Evolution in General, and that of Darwin, Goethe, and Lamarck in Particular* (New York: Appleton, 1868), p. 172. Quoted in John Vandermeer: *Reconstructing Biology: Genetics and Ecology in the New World Order* (New York: John Wiley and Sons, 1996), p. 21.

67 *"Genetics is, at last, like Germany"*: Steve Jones, *In the Blood: God, Genes and Destiny* (London: HarperCollins, 1996), p. vii.

69 *"To claim the possibility of some infallible scientific key"*: Isaiah Berlin, *The Sense of Reality: Studies in Ideas and Their History* (London: Chatto and Windus, 1996), p. 21.

70 *X-rays, injections, electrocution of the genitals:* Kevles, *In the Name of Eugenics*, p. 169.

71 *"the time was not right"*: Eugenics Society Records, file C108. Quoted in Kevles, *In the Name of Eugenics*, p. 252.

71 *"Eugenics is running the usual course of many new ideas"*: Ruth Hubbard and Elijah Wald, *Exploding the Gene Myth: How Genetic Information Is Produced and Manipulated by Scientists, Physicians, Employers, Insurance Companies, Educators and Law Enforcers* (Boston: Beacon Press, 1993), p. 15.

72 *Man's view of himself:* Philip Handler, ed., *Biology and the Future of Man* (Oxford: Oxford University Press, 1970), p. 926. Quoted in Kevles and Hood, *The Code of Codes*, pp. 287–88.

73 *The old eugenics:* Robert Sinsheimer, "The Prospect of Designed Genetic Change," *Engineering and Science* 32 (1969). Quoted in Kevles and Hood, *The Code of Codes*, pp. 289–90.

75 *I have suggested:* Troy Duster, *Backdoor to Eugenics* (New York, London: Routledge, 1990), p. 46.

78 *"In a world where each pair must be limited"*: Bentley Glass, "Science: Endless Horizons or Golden Age," *Science* 171 (1971), pp. 23–29. Quoted in Hubbard and Wald, *Exploding the Gene Myth*, p. 25.

78 *"Now the eugenic moral basis is this"*: G. K. Chesterton, *Eugenics and Other Evils* (London: Cassell and Co., 1922), p. 5.

79 *"the individual is an evanescent combination"*: E. O. Wilson, *On Human Nature* (London: Penguin Books, 1995), p. 197.

87 *"Does it really make a difference"*: Weir, Lawrence, and Fales, *Genes and Human Self-Knowledge*, p. 168.

88 *"If a child destined to have permanently low IQ"*: Daniel E. Koshland, Jr., "The Future of Biological Research: What Is Possible and What Is Ethical?", *MBL Science* 3 (1988–89), pp. 10–15. Quoted in Hubbard and Wald, *Exploding the Gene Myth*, pp. 115–16.

88 *"It is a short step"*: Ibid.

88 *"Idiots give birth to idiots"*: *The New York Times*, August 15, 1991, p. 1. Quoted in Kevles and Hood: *The Code of Codes*, p. 317.

88 *"Once we have left the garden of genetic innocence"*: Philip Kitcher, *The Lives to Come: The Genetic Revolution and Human Possibilities* (New York: Simon and Schuster, 1996), p. 204.

88 *"If prenatal testing for genetic diseases"*: Ibid., p. 198.

89 *Individual choices are not made in a social vacuum*: Ibid., p. 199.

89 *Utopian eugenics*: Ibid., p. 202.

90 *In 1989 a European Parliament committee*: Kevles and Hood, *The Code of Codes*, p. 320.

Chapter Five: The Mighty Gene

91 *Innocent of what?*: Little Bill, character played by Gene Hackman in Clint Eastwood's film *Unforgiven*, 1992.

91 *Behind the shifting face of personality*: Dorothy Nelkin and M. Susan Lindee, *The DNA Mystique: The Gene As Cultural Icon* (New York: W. H. Freeman and Co., 1995), p. 48.

91 *The secret is too plain*: John Ashbery, "Self-Portrait in a Convex Mirror," in *Selected Poems* (London: Penguin Books, 1994), p. 189.

93 *"If a person's genetic structure"*: Weir, Lawrence, and Fales, *Genes and Human Self-Knowledge*, p. 26.

96 *an intrinsic tension*: Ibid.

96 *"Our fundamental conception of ourselves as persons"*: Ibid., p. 28.

98 *It is especially ironic*: Nelkin and Lindee, *The DNA Mystique*, p. 126.

99 *"The products of the genome project"*: Timothy F. Murphy and Marc A. Lappé, eds., *Justice and the Human Genome Project* (Los Angeles: University of California Press, 1994), p. 162.

100 *"What is clear is that even a partial picture of the genetic landscape"*: Ibid.
101 *"The same tale, in different versions"*: Stephen Jay Gould, *The Mismeasure of Man* (London: Penguin Books, 1992), p. 20.
102 *"I was inspired to write this book"*: Ibid., p. 28.
102 *"We pass through this world but once"*: Ibid., pp. 28–29.
103 *"no large employer has been able to hire"*: Richard J. Herrnstein and Charles Murray, *The Bell Curve: Intelligence and Class Structure in American Life* (New York: Free Press, 1994), p. 85.
104 *"Putting it all together"*: Ibid., p. 91.
105 *"Try to envision what will happen"*: Ibid., p. 517.
105 *"Mounting evidence indicates"*: Ibid., p. 341.
105 *"Perhaps the ethical principles for not committing crimes"*: Ibid., pp. 240–41.
105 *"Inequality of endowments, including intelligence, is a reality"*: Ibid., pp. 551–52.
109 *"Over the past decade and a half"*: R. C. Lewontin, Steven Rose, and Leon J. Kamin, *Not in Our Genes: Biology, Ideology, and Human Nature* (New York: Pantheon Books, 1984), p. ix.
109 *"Sociobiology is yet another attempt"*: Ibid., p. 264.
110 *"We have reached a point"*: Derek Freeman, *Margaret Mead and the Heretic* (London: Penguin Books, 1996), p. 297.
114 *"For example, can a Darwinian understanding"*: Robert Wright, *The Moral Animal: Why We Are the Way We Are* (London: Abacus, 1994), p. 10.
114 *"in contrast to Freud"*: Ibid., p. 320.
115 *"If we are to recover social harmony and virtue"*: Matt Ridley, *The Origins of Virtue* (London: Viking Press, 1996), p. 264.
115 *For St. Augustine the source of social order*: Ibid.
117 *"A theory that explains everything"*: John Vandermeer, *Reconstructing Biology: Genetics and Ecology in the New World Order* (New York: John Wiley and Sons, 1996), p. 163.

Chapter Six: Seeds of Destruction

121 *"There is, of course, absurd suffering"*: Murphy and Lappé, *Justice and the Human Genome Project*, p. 11.
121 *"In the past fifty years"*: R. C. Lewontin, *The Doctrine of DNA: Biology as Ideology* (London: Penguin Books, 1993), pp. 42–43.
122 *"We contain within us the seeds of our own destruction"*: Levine and Suzuki, *The Secret of Life*, p. 91.
122 *A geneticist can work for years in a laboratory*: Robert Cook-Deegan, *The Gene Wars: Science, Politics and the Human Genome* (New York: W. W. Norton and Co., 1995), pp. 24–25.

124 *"the incredible eagerness with which scientists"*: Suzuki and Knudtson, *Genethics*, p. 138.

124 *"Like some prescientific shamanistic healer"*: Ibid.

125 *"My God, once we've identified"*: Levine and Suzuki, *The Secret of Life*, p. 31.

128 *The expense of genetic tests:* Murphy and Lappé, *Justice and the Human Genome Project*, p. 51.

128 *"The result will be that people who feel healthy"*: Weir, Lawrence, and Fales, *Genes and Human Self-Knowledge*, p. 29.

130 *"When the only option is to terminate a pregnancy"*: Kitcher, *The Lives to Come*, p. 84.

132 *"the father of 'client-centered counseling'"*: Cranor, *Are Genes Us?*, p. 151.

132 *"as clarifying and objectifying the client's own feelings"*: Ibid., p. 152.

132 *"Counselors will sometimes be the last and only possibility"*: Kitcher, *The Lives to Come*, p. 79.

133 *"In social terms, the mere existence of a screen"*: Duster, *Backdoor to Eugenics*, p. 79.

133 *"Genetic testing is encouraged by legal pressures"*: Kevles and Hood, *The Code of Codes*, p. 179.

133 *"A counselor should help a person who is at risk"*: Weir, Lawrence, and Fales, *Genes and Human Self-Knowledge*, p. 61.

133 *"There is no treatment or cure for HD"*: Ibid., p. 11.

135 *"There are many problems associated with the geneticization"*: Cranor, *Are Genes Us?*, p. 97.

135 *"In fact, molecular techniques should be understood"*: Ibid., p. 102.

136 *"Certainly, we hope at least to be able to foretell"*: Murphy and Lappé, *Justice and the Human Genome Project*, p. 7.

137 *"In a Darwinian world there is no longer any norm"*: Cranor, *Are Genes Us?*, p. 124.

137 *"a statement that disabled people shouldn't exist"*: Nelkin and Lindee, *The DNA Mystique*, p. 174.

137 *"people like me shouldn't exist"*: London: *Sunday Times*, January 1, 1995.

137 *"It would seem to make little sense"*: Kevles and Hood, *The Code of Codes*, p. 326.

138 *"permit the erosion of difference in favor of genetic uniformity"*: Murphy and Lappé, *Justice and the Human Genome Project*, p. 8.

Chapter Seven: The Spider

142 *Tolstoy, Shakespeare, Dostoevsky, Kafka, Nietzsche:* Berlin, *The Sense of Reality*, pp. 20–21.

142 *"an automatic orientation towards information"*: Wilson, *On Human Nature*, p. 169.

144 *"For in fact what is man in nature"*: Blaise Pascal, *Pensées*, trans. W. F. Trotter (London: Everyman, 1947), pp. 17–18.

145 *"I have no need of that hypothesis"*: Ian Stewart, *Does God Play Dice? The Mathematics of Chaos* (Oxford: Basil Blackwell, 1989), p. 10.

146 "importance itself, *like everything else that we treasure"*: Daniel C. Dennett, *Darwin's Dangerous Idea: Evolution and the Meaning of Life* (London: Penguin Books, 1996), p. 184.

147 *"just what we need in our attempt"*: Ibid., p. 521.

149 *"We feel that even when all possible scientific questions"*: Ludwig Wittgenstein, *Tractatus Logico-Philosophicus* (London: Routledge and Kegan Paul, 1951), p. 187.

149 *"Together they strike a fundamental blow"*: Dennett, *Darwin's Dangerous Idea*, p. 207.

150 *"A theory that explains everything"*: Vandermeer, *Reconstructing Biology*, p. 163.

151 *"Everything we know about molecular biology"*: Crick, *What Mad Pursuit*, p. 61.

152 *"Bach is precious"*: Dennett, *Darwin's Dangerous Idea*, p. 512.

152 *"only adds to the excitement and mystery and awe"*: Richard Feynman, *What Do You Care What Other People Think? Further Adventures of a Curious Character* (London: Unwin, 1990), p. 11.

152 *"There is grandeur in this view of life"*: Charles Darwin, *On the Origin of Species* (1859; reprint, Cambridge, Mass.: Harvard University Press, 1994), p. 490.

154 *"that scientific truth has human illusion as its regular by-product"*: Roger Scruton, *An Intelligent Person's Guide to Philosophy* (London: Duckworth, 1996), p. 8.

155 *"As our latest attempt at dropping some moral anchor"*: Howard Kaye, "Are We the Sum of Our Genes?," *Wilson Quarterly XVI*, pp. 77–84. Quoted in Cook-Deegan, *The Gene Wars*, p. 254.

155 *"This combination of vast power"*: Cranor, *Are Genes Us?*, p. 173.

155 *"our social mores and institutions"*: Murphy and Lappé, *Justice and the Human Genome Project*, p. 5.

155 *"This means that in science"*: Jones, *In the Blood*, p. xvi.

156 *"The Law . . . makes it clear that from observation statements"*: Peter Medawar, *The Limits of Science* (Oxford: Oxford University Press, 1984), pp. 81–82.

157 *"The concept of free choice"*: Weir, Lawrence, and Fales, *Genes and Human Self-Knowledge*, p. 167.

158 *"The first, that we not so place our felicity"*: Francis Bacon, *De Dignitate et*

Augmentis Scientiarum (trans. Wats, 1674). Quoted in Medawar, *The Limits of Science*, p. 66.

160 *"The unknown and prodigious are drugs"*: Edward O. Wilson, *The Diversity of Life* (London: Allen Lane, Penguin Press, 1993), p. 7.

161 *"This horrifies me, although I cannot say why"*: William Cookson, *The Gene: Adventures in the Genome Jungle* (London: Aurum Press, 1994), p. 105.

163 *"Even those who do not believe in God"*: Weir, Lawrence, and Fales, *Genes and Human Self-Knowledge*, p. 243.

163 *"Among such people"*: Ibid.

163 *"operate in a world in which natural assets and liabilities"*: Murphy and Lappé, *Justice and the Human Genome Project*, p. 145.

165 *"An intellect which at any given moment"*: Pierre Simon de Laplace, *A Philosophical Essay on Probabilities* (New York: Dover, 1951). Quoted in Stewart, *Does God Play Dice?*, p. 10.

165 *"The ontological implication of human genetics"*: Weir, Lawrence, and Fales: *Genes and Human Self-Knowledge*, p. 38.

165 *"because any theory about it is compatible"*: Ibid., p. 48.

165 *"What matters is whether a person"*: Ibid.

166 *"Ultimately the impact of molecular biology"*: Levine and Suzuki, *The Secret of Life*, p. 250.

166 *"One of our ideological preconceptions in this country"*: Ibid., pp. 248–49.

169 *"Because our world is not the same as Othello's world"*: Huxley, *Brave New World*, pp. 201–2.

170 *"The ideology of individualism is a powerful friend"*: Kitcher, *The Lives to Come*, p. 268.

170 *"Today's biological organism bears little resemblance"*: Evelyn Fox Keller, *Refiguring Life: Metaphors of Twentieth-Century Biology* (New York: Columbia University Press, 1995), pp. 117–18.

170 *"In short, the science is wonderful"*: Tudge, *The Engineer in the Garden*, p. 386.

bibliography

Berlin, Isaiah. *The Sense of Reality: Studies in Ideas and Their History*. London: Chatto and Windus, 1996.

Bishop, Jerry E., and Waldholz, Michael. *Genome: The Story of Our Astonishing Attempt to Map All the Genes in the Human Body*. New York: Simon and Schuster, 1991.

Bodmer, Walter, and McKie, Robin. *The Book of Man: The Quest to Discover Our Genetic Heritage*. London: Abacus, 1995.

British Medical Association. *Our Genetic Future: The Science and Ethics of Genetic Technology*. Oxford: Oxford University Press, 1992.

Bud, Robert. *The Uses of Life: a History of Biotechnology*. Cambridge: Cambridge University Press, 1993.

Chesterton, G. K. *Eugenics and Other Evils*. London: Cassell and Co., 1922.

Cook-Deegan, Robert. *The Gene Wars: Science, Politics and the Human Genome*. New York, London: W. W. Norton and Co., 1995.

Cookson, William. *The Gene Hunters: Adventures in the Genome Jungle*. London: Aurum Press, 1994.

Cranor, Carl F., ed. *Are Genes Us? The Social Consequences of the New Genetics*. New Brunswick, N.J.: Rutgers University Press, 1994.

Crick, Francis. *The Astonishing Hypothesis: The Scientific Search for the Soul*. London: Touchstone Books, 1995.

———. *What Mad Pursuit: A Personal View of Scientific Discovery*. London: Penguin Books, 1990.

Darwin, Charles. *On the Origin of Species*. Cambridge, Mass.: Harvard University Press, 1964.

Davis, Bernard D. *The Genetic Revolution: Scientific Prospects and Human Perceptions*. Baltimore: Johns Hopkins University Press, 1992.

Dennett, Daniel. *Darwin's Dangerous Idea: Evolution and the Meaning of Life.* London: Penguin Books, 1996.

Dixon, Patrick. *The Genetic Revolution.* Eastbourne: Kingsway Publications, 1993.

Duster, Troy. *Backdoor to Eugenics.* New York, London: Routledge, 1990.

Eldredge, Niles. *Reinventing Darwin: The Great Evolutionary Debate.* London: Phoenix, 1996.

Fox Keller, Evelyn. *Refiguring Life: Metaphors of Twentieth Century Biology.* New York: Columbia University Press, 1995.

Freeman, Derek. *Margaret Mead and the Heretic.* London: Penguin Books, 1996.

Galton, Francis. *Hereditary Genius: An Inquiry into Its Laws and Consequences.* London: Macmillan, 1869.

Glyn-Jones, Anne. *Holding Up a Mirror: How Civilizations Decline.* London: Century, 1996.

Gould, Stephen Jay. *The Individual in Darwin's World.* Edinburgh: Edinburgh University Press, 1990.

———. *The Mismeasure of Man.* London: Penguin Books, 1992.

Gribbin, John. *In Search of the Double Helix: Quantum Physics and Life.* London: Penguin Books, 1995.

Hall, Stephen. *Invisible Frontiers: The Race to Synthesize the Human Gene.* London: Sidgwick and Jackson, 1988.

Hamer, Dean, and Peter Copeland. *The Science of Desire: The Search for the Gay Gene and the Biology of Behavior.* New York: Simon and Schuster, 1994.

Herrnstein, Richard J., and Charles Murray. *The Bell Curve: Intelligence and Class Structure in American Life.* New York: Free Press, 1994.

Hubbard, Ruth, and Elijah Wald. *Exploding the Gene Myth: How Genetic Information Is Produced and Manipulated by Scientists, Physicians, Employers, Insurance Companies, Educators and Law Enforcers.* Boston: Beacon Press, 1993.

Huxley, Aldous. *Brave New World.* London: Flamingo, 1994.

Jones, Steve. *In the Blood: God, Genes and Destiny.* London: HarperCollins, 1996.

———. *The Language of the Genes: Biology, History and the Evolutionary Future.* London: Flamingo, 1994.

Kevles, Daniel J. *In the Name of Eugenics: Genetics and the Uses of Human Heredity*. Cambridge, Mass.: Harvard University Press, 1995.

Kevles, Daniel J., and Leroy Hood, eds. *The Code of Codes: Scientific and Social Issues in the Human Genome Project*. Cambridge, Mass.: Harvard University Press, 1993.

Kitcher, Philip. *The Lives to Come: The Genetic Revolution and Human Possibilities*. New York: Simon and Schuster, 1996.

Levine, Joseph, and David Suzuki. *The Secret of Life: Redesigning the Living World*. Boston: WGBH Educational Foundation, 1993.

Lewin, Roger. *Complexity: Life at the Edge of Chaos*. London: J. M. Dent, 1993.

Lewontin, R. C. *The Doctrine of DNA: Biology As Ideology*. London: Penguin Books, 1993.

Lewontin, R. C., Steven Rose, and Leon J. Kamin. *Not in Our Genes: Biology, Ideology and Human Nature*. New York: Pantheon Books, 1984.

Lyon, Jeff, and Peter Gorner. *Altered Fates: Gene Therapy and the Retooling of Human Life*. New York: W. W. Norton & Co., 1995.

McCarty, Maclyn. *The Transforming Principle: Discovering That Genes Are Made of DNA*. New York: W. W. Norton & Co., 1985.

Medawar, Peter. *The Limits of Science*. Oxford: Oxford University Press, 1986.

Murphy, Michael P., and Luke A. J. O'Neill, eds. *What Is Life? The Next Fifty Years: Speculations on the Future of Biology*. Cambridge: Cambridge University Press, 1995.

Murphy, Timothy F., and Marc A. Lappé, eds. *Justice and the Human Genome Project*. Los Angeles: University of California Press, 1994.

Nelkin, Dorothy, and Susan M. Lindee. *The DNA Mystique: The Gene As Cultural Icon*. New York: W. H. Freeman & Co., 1995.

Olby, Robert. *The Path to the Double Helix: The Discovery of DNA*. New York: Dover Publications, 1994.

Rattray Taylor, Gordon. *The Biological Time-Bomb*. London: Thames & Hudson, 1968.

Ridley, Matt. *The Origins of Virtue*. London: Viking, 1996.

Rose, Steven, with Sarah Bullock. *The Chemistry of Life*. London: Penguin Books, 1991.

Ruse, Michael, ed. *But Is It Science? The Philosophical Question in the Creation/ Evolution Controversy.* New York: Prometheus Books, 1996.

Schrodinger, Erwin. *What Is Life?* Cambridge: Cambridge University Press, 1967.

Scruton, Roger. *An Intelligent Person's Guide to Philosophy.* London: Duckworth, 1996.

Shaw, George Bernard. *Man and Superman.* London: Penguin Books, 1957.

Suzuki, David, and Peter Knudtson. *Genethics: The Clash Between the New Genetics and Human Values.* Cambridge, Mass.: Harvard University Press, 1990.

Thompson, Larry. *Correcting the Code: Inventing the Genetic Cure for the Human Body.* New York: Simon and Schuster, 1994.

Tudge, Colin. *The Engineer in the Garden. Genes and Genetics: From the Idea of Heredity to the Creation of Life.* London: Jonathan Cape, 1993.

Vandermeer, John. *Reconstructing Biology: Genetics and Ecology in the New World Order.* New York: John Wiley and Sons, 1996.

Wallace, Bruce. *The Search for the Gene.* Ithaca, N.Y.: Cornell University Press, 1992.

Watson, James D. *The Double Helix: A Personal Account of the Discovery of the Structure of DNA.* London: Weidenfeld and Nicolson, 1968.

Watson, Lyall. *Supernature: A Natural History of the Supernatural.* London: Sceptre, 1974.

Weir, Robert F., Susan C. Lawrence, and Evan Fales, eds. *Genes and Human Self-Knowledge: Historical and Philosophical Reflections on Modern Genetics.* Iowa City: University of Iowa Press, 1994.

Wilkie, Tom. *Perilous Knowledge: The Human Genome Project and Its Implications.* London: Faber and Faber, 1994.

Wilson, Edward O. *Biophilia.* Cambridge, Mass.: Harvard University Press, 1994.

———. *The Diversity of Life.* London: Penguin Books, 1993.

———. *On Human Nature.* London: Penguin Books, 1995.

Wright, Robert. *The Moral Animal: Why We Are the Way We Are.* London: Abacus, 1996.

Index

counselor, role of, 132
creationism:
    vs. evolution, 26–27, 39, 40, 150
    and patented body parts, 39
creation science, 27
Crick, Francis, 149, 153, 164
    and DNA, 1–2, 6, 7, 34, 74, 91, 151
    and ethical dilemma, 29
    and molecular psychology, 22
crimes:
    and dysgenic pressure, 105
    genetic defense for, 94–96
    in the name of eugenics, 44, 48–49, 67
criminal gene, 23, 94–96, 100
criminal justice:
    and capital punishment, 61
    and inequality, 101
    and responsibility, 94–98
cystic fibrosis, 75, 77, 130, 135

Darrow, Clarence, 27
Darwin, Charles, 46, 53, 56, 149, 152–53
Darwinism, see evolution
Davenport, Charles B., 61, 63
Davies, Kay, 135, 138
Davis, Bernard D., 41
Dawkins, Richard, 92, 115
DDT, 29–30
death:
    inevitability of, 11
    perfect, 123, 125
    and probabilities, 14
democracy:
    vs. aristocracy, 49, 50
    as the end of history, 167
    equality in, 100, 109, 166–67
    ethics in, 37–38, 49, 61
    freedom in, 132, 134, 157, 167, 169
    intelligence in, 104
    neutrality in, 167
Dennett, Daniel, 146–47, 149–50, 152, 154, 158
Denno, Deborah, 96
depression, 15–16, 45
determinism:
    biological, 102, 104, 109
    free will vs., 165–66
    genetic, 170
Diana, Princess of Wales, 49
disease:
    abortion as prevention of, 16–17, 81, 84–85, 89, 120, 129–31

conquest of, 138, 139, 140
crops resistant to, 24
definitions of, 16, 128
environmental factors in, 12–13
as error, 139–40
and genetics, see medical genetics
and human differences, 77, 100
identifying predisposition for, 125
knowledge as power over, 122, 138
and life expectancy, 11
mental illnesses as, 15–16, 22
as moral dimension, 74–76
multigene, 12–14
and probabilities, 14, 125
and race, 77
rare and difficult, 123
and recessive alleles, 125
right use of, 121, 139
risk factors for, 12–13
science of, 123
single–gene, 12, 75–77, 119–20, 125
as universal, 128–29
see also specific diseases
divination, 136
DNA (deoxyribonucleic acid):
    and abortion, 131
    and aging, 11
    and cancer, 19, 122
    and designer babies, 17
    and eugenics, 45
    and evolution, 6, 32, 34, 36
    and genetics, 3, 9, 47–48, 91, 125
    and Human Genome Project, 10, 99
    and insurance companies, 22–23
    as magic, 91–92
    and mental illness, 47
    moratorium on research in, 31–36
    and privacy, 127
    and probabilities, 14, 80
    recombinant, 31–36, 73
    similarities in, 98–99
    structure of, 2–3, 6, 34, 74, 91
    as total being, 79, 151
    and transgenics, 24
    and YACs, 20
Dolly, see cloning
Down's syndrome, 71, 81, 130, 134–35
dualism vs. monism, 65
Dugdale, Richard, 60
Duster, Troy, 78, 133
dysgeny, 57, 66, 72, 105

education, 71, 85, 104
Einstein, Albert, 8, 34, 68, 159
embryos, preselection of, 84–85
employment, and testing, 103–4
Enlightenment, 50–52, 100
environment:
  alien, ability to prosper in, 36
  and anxiety, 41–42
  benign, 112
  vs. biology, *see* nature vs. nurture
  changes in, 56–57
  chemical traces in, 29–30
  and disease, 12–13
  and evolution, 146
  and Green party, 41
  and homosexuality, 83
  and human character, 98
  and life, 13
  maximization of, 111
  as other people's genes, 94, 114
  overpopulation in, 78
  and transvestitism, 93–94 ' '
epiphenomena, 145–46
equality of opportunity, 98, 99, 100, 104,
  108, 109, 166–67
ethics:
  and abortion, 38–39
  and anxiety, 37–42
  and biology, 29, 31
  of cloning, 37–38, 156
  in a democracy, 37–38, 49, 61
  and meaning, 157, 161–62
  and moral apocalypse, 29–30
  vs. political rhetoric, 38
  and technology, 38, 62–63, 127
  use of term, 37
eugenics, 44–64, 65–90
  and abortion, 80–81, 88–89, 134–35
  and biology, 45–47, 62, 67, 73
  central principle of, 56–57, 63
  consumer, 87
  and deviations from the norm, 86–
    87
  and disease, 74–76
  of free market, 84, 86, 87, 89, 157
  and handicaps, 135, 137, 138
  and heredity, 59, 61
  and homosexuality, 83–86
  interventions of, 88, 89
  and mental illness, 45, 47–48, 72
  modern, 53–64
  and molecular biology, 74

  moral issues in, 44–49, 59–60, 64, 70–72,
    73–74, 79–82, 90
  philosophical issues in, 50–52
  and politics, 60, 61, 70–72
  and population control, 78–79
  private decisions in, 82, 84, 86
  psychological issues in, 52–53, 58
  reform, 63
  in selecting life partners, 49, 54, 82
  and statistics, 56, 60
  and sterilization, 47, 61, 70, 71, 80, 81
  and substance abuse, 45, 47, 129
  use of word, 49, 80–81
  utopian, 89, 126
evolution:
  and abortion, 80
  adaptation in, 113
  anxiety produced by, 26, 28
  and behavior, 113
  competition in, 111
  and connectedness of all life, 98
  vs. creation theory, 26–27, 39, 40, 150
  and DNA, 6, 32, 34, 36
  and dysgeny, 57, 66, 72
  and environmental pressures, 146
  and game theory, 110–14
  group selection in, 112
  guiding our own, 72–73
  and heredity, 56, 59
  and human sexual behavior, 22
  and integrity of the species, 49, 54, 56–60,
    67
  and logotaxis, 142–43
  and monism, 67
  and natural selection, 113, 145
  and nature, 53, 72–73
  process of, 56–57
  and religion, 53, 92
  and society, 137
  and subconscious, 115
  as Theory of Everything, 116–17, 150
  and tinkering with the gene pool, 21
evolutionary psychology, 21–22, 110,
  113–14, 116, 147–48, 153–54

Ferguson, Sarah, 49–50
fetus, and personhood, 140–41
Feynman, Richard, 152
fight or flight, 95
Fischer, Eugene, 48
Fleck, Leonard M., 163
food, genetically engineered, 36, 40, 41–42

political issues: *(cont.)*
  balancing of forces in, 38
  capital punishment, 61
  education, 71, 85
  ethics in, 38, 61, 71, 127
  eugenics, 60, 61, 70–72
  in genetic studies, 23, 42
  in genocentrism, 92, 115
  private vs. public, 71
  socialist Utopia, 60
  sociobiology, 115
polygenic disorders, 12–14
population control, 78–79
population genetics, 56
pornography, 82
Prayer to Ask God for the Right Use of Sickness, A (Pascal), 121, 139
preimplantation diagnosis, 17
prejudice, *see* race
prenatal testing, 88, 89
  and abortion, 120, 127, 130–31, 137
  and designer babies, 17, 81–82, 84–85
  and public policy, 128
Principles of Human Heredity and Race Hygiene (Fischer), 48
Prisoner's Dilemma, 111–13
probability:
  and medical genetics, 13–15, 125
  and personhood, 80
  and prenatal testing, 17
proteins, 9, 18, 91
psychology:
  and eugenics, 52–53, 58
  evolutionary, 21–22, 110, 113–14, 116, 147–48, 153–54
  molecular, 22

quantum theory, 2, 4, 8
Quinlan, Karen Ann, 71

race:
  and biological determinism, 102, 104, 109
  and disease, 77
  and education, 104
  and eugenics, 45, 48
  and intelligence, 101–9
  and monism, 66–67
Ravetz, Jerome, 34
recessive alleles, 125
recessive single-gene disorders, 125
recombinant DNA, 31–36, 73
Reich, Robert, 105

relativity theory, 8
religion:
  and absolute of the soul, 116, 151–53
  decline of, 116
  and Enlightenment, 50–52
  and evolution, 53, 92
  and genes replacing soul, 147
  and knowledge, 144–45, 148–49
  as last refuge, 151–54, 171
  and meaning, 162
  and mind as inner sanctum, 150
  and molecular basis of life, 6
  and morality, 59, 70
  and patented body parts, 39
  and redemptive suffering, 139–40
  and sanctity of human life, 40, 42, 158
  vs. science, 39, 40, 145, 148–49, 153
  and Spanish Inquisition, 44–49, 62
reproduction:
  vs. abortion, *see* abortion
  control of, 63, 73–74, 76, 78–79
  as end goal of life, 11
  and genetics, 5
  and mental illness, 72
  and qualitative judgments, 58
  state intervention in, 84, 88
Ridley, Matt, 115–16
Rifkin, Jeremy, 40–42, 88
Rogers, Carl, 132
Rose, Steven, 109
Rous, Peyton, 19
Rousseau, Jean-Jacques, 115
Rudin, Ernst, 47
Ruse, Michael, 165
Russell, Bertrand, 59, 71

sarc gene, 19
schizophrenia, 15–16, 45, 47
science:
  and anxiety, 26–43
  and behavior, 96
  as black box, 3–4, 91
  contribution of, 120–21
  controls in, 36
  creation, 27
  cultural bias in, 68, 102, 161–62
  and democracy, 100
  and Enlightenment, 50–51
  errors of, 159, 162
  and genetics, 5, 46–47, 123
  as global destroyer, 28, 33
  good, in evil regime, 46–47, 68, 167

and history, 68–70
in the human realm, 68–70, 164, 166
as illusion, 154, 156, 159
impact of, 5–6, 34–35, 68–69, 121, 149–58
impermanence of, 159
and inequality, 100
and lay people, 3–6
and logotaxis, 145, 148, 161
mistrust of, 37, 40–41, 150–51, 156–67, 172
and morality, 59, 61, 69–70, 92, 96, 154–55, 168
moratorium on research in, 33–36
paradigm shift in, 114
persuasive promise of, 42
and radical reductionism, 69–70
as rationale for slaughter, 70
vs. religion, 39, 40, 145, 148–49, 153
responsibility in, 34, 67
as technology, 31–36, 73–74, 159, 170–71
and truth, 68, 73, 164
writers of, 5
and yuk factor, 162–63
see also specific sciences
science fiction, 160
scientism, 156
Scopes trial, 26–27, 40, 82
Scruton, Roger, 154, 156, 158
Secret of Life, The (Levine and Suzuki), 125
selfish gene, 110, 112, 113, 115
sex cells, and germ-line therapy, 21
Shakespeare, Tom, 137
Shakespeare, William, 86, 169
Shaw, George Bernard, 59–60, 71
shortness, as deviation from norm, 86, 136
sickle-cell anemia, 12, 21, 75–77
sickness, *see* disease
Silent Spring (Carson), 29–30
Singapore, eugenics in, 88
Sinsheimer, Robert, 73
Sistine Chapel, 153
slaves, 39, 49, 50
Smith, Adam, 109
Smith, John Maynard, 111
smoking, 126, 138
Social Darwinism, 60
social justice, 94–98
social structure:
    and capital punishment, 61
    children in, 63, 85, 87
    consumerism in, 87
    democracy in, 49, 50, 134

divisiveness in, 138
and education, 71, 85, 104
individual vs. collective good in, 132, 169
instability in, 101
intelligence in, 103–9
and justice, 163, 166
and knowledge, 76, 96
masses vs. elite in, 57–58, 60
and mental retardation, 89
normality in, 83, 85–86, 93, 135, 136
and prenatal diagnosis, 128
public vs. private decisions in, 82, 85
reduction of, 115, 156
sociobiology, 108–9, 111, 115, 153–54
sociology, 110, 127–28
Socrates, 116
Solzhenitsyn, Aleksandr, 162
soul:
    absolute of, 116
    as last refuge, 151–54, 171
    replaced by genes, 147
Soviet Union, and socialist Utopia, 60
space travel, 8–9
Spanish Inquisition, 44–49, 62
Spencer, Herbert, 60
spider, as knowledge, v, 147, 167
standard distribution, 55
Star Wars, 160
statistics:
    bell curve in, 55, 106–7
    and the big picture, 57
    and eugenics, 56, 60
    and genetics, 54–56, 60, 94
    and the individual, 106–7
    and moral issues, 55–57, 60, 80
    and nature vs. nurture, 94
    and probability, 14–15, 17, 80, 125
    standard distribution of, 55
sterilization, 47, 61, 70, 71, 80, 81
subconscious, 115
Supreme Court, U.S., 103
Suzuki, David, 124–25, 166
SV40 (monkey tumor virus), 33
Szilard, Leo, 34

taboos, 162–63
Taylor, Gordon Rattray, 29
Tay-Sachs disease, 76–77, 81
technology:
    and democracy, 100
    and ethical issues, 38, 62–63, 127
    and free market, 167–68

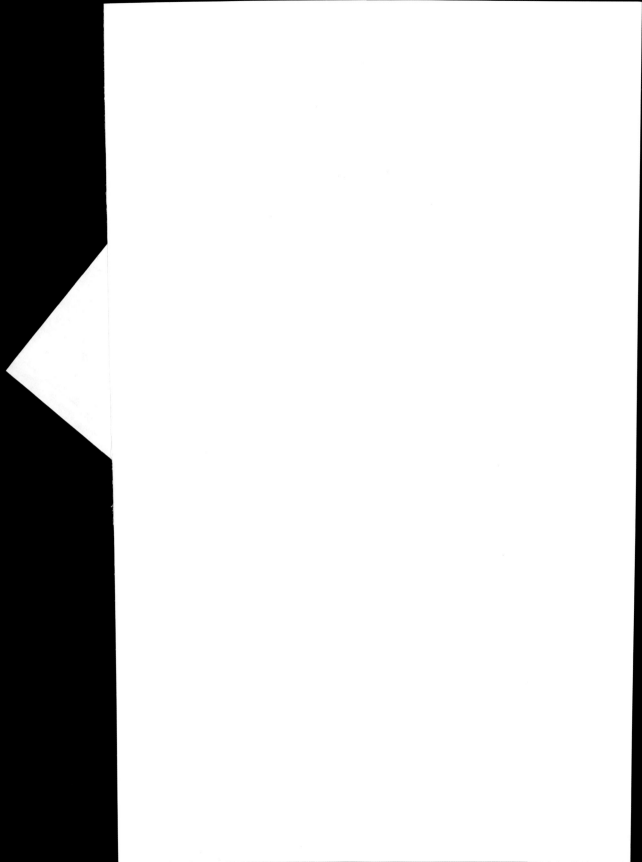

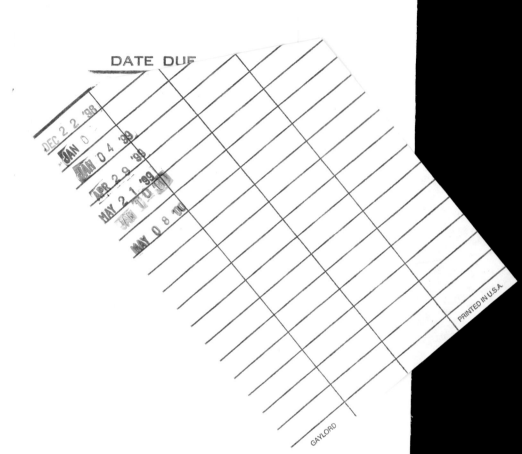